Principles of Surface Chemistry

FUNDAMENTAL TOPICS IN PHYSICAL CHEMISTRY

Editor: **Harold S. Johnston,** *University of California, Berkeley*

HANNAY *Solid-state Chemistry*

MYERS *Molecular Magnetism and Magnetic Resonance Spectroscopy*

SOMORJAI *Principles of Surface Chemistry*

STRAUSS *Quantum Mechanics: An Introduction*

WESTON and SCHWARZ *Chemical Kinetics*

G. A. SOMORJAI

Principles

of

Surface

Chemistry

PRENTICE-HALL, INC.

Englewood Cliffs, New Jersey

FUNDAMENTAL TOPICS IN PHYSICAL CHEMISTRY
Editor: Harold S. Johnston

To my parents

Printed in the United States of America
Library of Congress Catalog Card No. 79-172890
ISBN: 0-13-710608-4

10
9
8
7
6
5
4
3
2
1

PRENTICE-HALL INTERNATIONAL, INC., *London*
PRENTICE-HALL OF AUSTRALIA, PTY. LTD., *Sydney*
PRENTICE-HALL OF CANADA, LTD., *Toronto*
PRENTICE-HALL OF INDIA PRIVATE LTD., *New Delhi*
PRENTICE-HALL OF JAPAN, INC., *Tokyo*

Contents

v

2. THERMODYNAMICS OF SURFACES, 52

3. DYNAMICS OF SURFACE ATOMS, 82

Foreword

THE EMPHASIS IN PHYSICAL-CHEMICAL RESEARCH has shifted over the past generation from the macroscopic view of thermodynamics, electrochemistry, and empirical kinetics to the molecular view of these and other areas. Statistical mechanics has provided a practical method for the evaluation of certain types of thermodynamic data. The microscopic viewpoint of solid-state physics has injected new intellectual vigor into electrochemistry, surface chemistry, and heterogeneous catalysis. Recent developments in molecular beams have turned chemical kinetics into one of the most active fields of physical chemistry. Thus most areas of physical chemistry now present two aspects, the old and the new; and this dichotomy is reflected in the teaching of physical chemistry.

The usual curriculum in chemistry includes a year's course in physical chemistry given during the sophomore or junior year. In some cases, teachers of this course try to retain all the old physical chemistry and to introduce all the new topics as well; then the sheer volume of material is overwhelming. In other cases, an effort is made to cover all the material from the modern point of view, starting with quantum mechanics; this approach tends to neglect (or treat in only a superficial way) complicated molecules, solutions, and the liquid state. It ought to be recognized that physical chemistry is not a body of knowledge that must be covered but rather a set of methods for predicting chemical events. Each of the methods should be understood. It is desirable to present chemical thermodynamics by the macroscopic approach, lest this method of science be omitted from the student's education. Next, it is imperative to teach the student a small amount of rigorous, Schrödinger-type quantum mechanics, with some—but not all—applications to molecular structure and molecular spectroscopy. The student of physical chemistry must also receive an introduction to statistical mechanics but not a survey of all its applications. With a rigorous, though limited, introduction to thermodynamics, quantum mechanics, and statistical mechanics, the student can readily be introduced to the methods and viewpoints of chemical kinetics, solid-state chemistry, and other areas of physical chemistry.

The ideal textbook in physical chemistry might contain a condensed, clear exposition of each *method of prediction* in chemistry, with extra reading material and references. Such a book would be long, and a course based on it inflexible. It therefore seems desirable to have available a series of relatively brief texts that can help serve the same purpose and that can give the teacher greater flexibility in his choice of teaching materials.

This then is the aim of the series Fundamental Topics in Physical Chemistry. Each author has been urged to make his goal student insight into one basic area in physical chemistry.

Harold S. Johnston

Preface

THIS BOOK is intended primarily for students who have had introductory courses in physical chemistry. It is my hope that upper division and graduate students as well as researchers working in many areas of chemistry, biochemistry, engineering sciences, and physics will find it a useful introduction to surface science.

Surface chemistry finds application in many technologies—for example, food and wood; in a great number of industries—oil, polymer, electronics, computers, mining, and ceramic, to name a few; and in space exploration. This great need has been brought about by rapid development of several areas of applied surface science, among them, heterogeneous catalysis, photography, and colloid science.

Early surface studies concentrated on the physical chemistry of the adsorbed layer since gas free, clean surfaces could not be prepared easily. The adsorption thermodynamics, and changes of electrical properties during chemical changes at the interface were the main subjects of research studies. There are several excellent books (most of them are referred to in the text) that treat some of these properties of the adsorbed layer in depth.

In the past, studies of the physical-chemical properties of gas-free surfaces were hampered by the scarcity of experimental tools that provide information about the surface atoms without containing a great deal of information about atoms in the bulk. A typical solid, like silver, has a bulk density of about 5.9×10^{22} atoms/cm³. The surface density of atoms can be estimated to be approximately the $\frac{2}{3}$ power of the volume density that is equal to 1.5×10^{15} atoms/cm². Thus in any experiment that is aimed at studying the surface, one attempts to investigate the properties of $\approx 10^{15}$ atoms/cm² in the background of a much larger concentration of bulk atoms. Surface chemists, therefore, were forced to prepare and carry out measurements on samples of high surface/volume ratio, i.e., small particles. Many ingenious experiments were carried out to determine the particle sizes accurately, to measure the amount of gas adsorbed on the surface of the particles, and to monitor the rates of chemical reactions that take place. In spite of the very large efforts of generations of surface scientists, there were serious limitations to the investigations of some of the fundamental properties of clean and gas-covered surfaces through studies of such dispersed systems of small particles.

Since the middle 1950's, there have been several technological developments that focused attention on crystal surfaces and on studies of their physical-chemical properties. Semiconductor devices with large surface to volume ratios necessitated a better understanding of the structural and electrical properties of clean surfaces. In addition, our efforts in space exploration have catalyzed the development of ultra high vacuum ($< 10^{-8}$ torr) technology which, in turn, made it possible to investigate crystal surfaces that are initially free of adsorbed gases.

As a consequence, the physical-chemical properties of surfaces are being uncovered at an unprecedented rate. The structure, thermodynamics, dynamics and electrical properties of the clean surface and of the adsorbed layer are presently the subjects of intense investigations by surface scientists and the discussion of these properties is the subject of this book.

I have concentrated on discussing the properties of mostly solid surfaces for several reasons. Our understanding of the various physical-chemical properties of solids is much better at present than that of the physical-chemical properties of liquids. Most of my research is with solid surfaces and discussion of their properties is closer to my scientific interest. Crystal surfaces are structurally well defined and can be prepared with relative ease. Finally, because of the availability of well characterized solid surfaces, most of the important advances in surface science in recent years have come from studies of solid surfaces.

In order to partially compensate for the lack of adequate training of a chemistry student in solid state chemistry, two brief sections (3.1 and 4.3) on the vibration and electrical properties of atoms in solids were inserted to facilitate discussion of the dynamical and electrical properties of surface atoms. Fortunately, the recent publication of a book on solid state chemistry in this series by N. B. Hannay has made broader discussion of solid state properties unnecessary.

In organizing the material, I have decided to emphasize those concepts and models of surface chemistry that are strongly supported by experiments (the exceptions are the theoretical estimates in sections 2.5 and 3.2). The book is divided into five major chapters. The first four concentrate on the physical chemical properties of "clean" surfaces, i.e., the properties of surfaces in the absence of adsorbed atoms. The first chapter discusses the structure of surfaces. Knowledge of the atomic surface structure is all important in interpreting any other surface property. The second chapter treats surface thermodynamics. The third chapter covers the dynamics of surface atoms and the electrical properties of surfaces are covered by the fourth chapter. I have attempted, while discussing these properties, to mention most of those experimental techniques that are being used in modern surface studies.

In the fifth and longest chapter, the interaction of gases with surfaces and the properties of the adsorbed layer are discussed. Recent advances in molecular beam techniques forwarn of great progress in the understanding of the dynamics of gas-surface interactions. In treating the properties of the adsorbed layer, the same order that was used to discuss the properties of clean surfaces is followed. The structure of the adsorbed layers is discussed first. The thermodynamics of weakly adsorbed and chemisorbed layers has been treated in many monographs and review papers so that I have restricted the discussion to that of principles and referred the interested reader to the voluminous literature. There is little information available at present on the dynamics of adsorbed layers. The discussion of the changes of electrical properties at the surface upon adsorption is followed by the discussion of the surface chemical bond that is the last section of this chapter.

The majority of the problems at the end of each chapter require that the student consult the original research papers in the literature for finding the solution. My purpose was to acquaint the student with some of the applications of the surface chemical concepts and with some of the important controversies that are so much part of any dynamic and developing field of science.

There are several important topics of surface chemistry that are not discussed in the book. Among them are some of the changes of phase, vaporization, field evaporation, condensation and crystal growth. These were left out mainly to limit the size of the text. The optical properties of surfaces are discussed only briefly (sections 5.4.4 and 5.6.9). A more detailed discussion did not appear warranted at this time since there are excellent reviews on the application of infrared techniques while other optical methods (ellipsometry, electron spin resonance, etc.) are in the midst of rapid development. I would have liked to discuss some of the important surface reactions, oxidation or the photographic process, for example. A discussion of the surface structural changes, the thermo-dynamics, dynamics and the changes of the electrical properties during the surface reaction would have been instructive. Unfortunately, none of the surface chemical reactions have been studied with all four of these properties in mind.

My sincere thanks go to the Guggenheim Foundation whose Fellowship made it possible to start and complete a considerable portion of this book while I was on sabbatical leave at Cambridge University. I am grateful for the kindness and hospitality of Professor J. W. Linnett, and I would like to acknowledge the hospitality extended to me by the staff of the Physical Chemistry Department, the Master and the fellows of Emmanuel College while I was visiting in Cambridge. I am grateful to my colleagues, Professor Harold S. Johnston and Professor G. Heiland for careful reading of the manuscript and many helpful suggestions, and to Professors Ronald Herm, George Jura, Richard Powell, and John Prausnitz for their helpful comments. I am grateful to many of my students and associates: Dr. Marlo Martin, Mr. L. A. West, Mr. J. Lira-Olivares, Mr. John Dancz, Miss M. Teng and Mr. R. Palmeri for critical reading and many suggestions concerning the manuscript. Many thanks to Mr. Charles Pezzotti and Mrs. Claudia Redwood for their painstaking work with the proofs.

I have had the good fortune to associate with one of the outstanding chemistry faculties. I would like to thank my colleagues in the College of Chemistry at the Berkeley campus of the University of California for the stimulation and instruction I have received from them.

Finally, I wish to acknowledge the support of the U.S. Atomic Energy Commission through the Inorganic Materials Research Division of the Lawrence Berkeley Laboratory for my research that has contributed directly or indirectly to this work.

Gabor A. Somorjai

List of Symbols

$\mathbf{T}$	translation vector
d	interrow distance
h, k, l	Miller indices
$\mathbf{G}$	reciprocal lattice vector
λ	wavelength
m	mass
E	total energy
v	velocity
h	Planck's constant
k	wave vector
ω	angular frequency
P_{eq}	equilibrium pressure
R	gas constant
n	integer
A	amplitude
M	atomic number
E^0	total energy per atom
E^S	total surface energy per unit area
$\mathscr{A}$	surface area

S	total entropy
S^0	total entropy per atom
S^s	surface entropy per unit area
A^s	surface work content per unit area
G^s	surface free energy per unit area
H^s	surface enthalpy per unit area
W^s	surface work
γ	surface tension
V	volume
r	radius
γ^0	surface tension at $T = 0°K$
C_P^S	surface heat capacity per unit area
μ_i	chemical potential of ith component
c_i^s	surface concentration of ith component per unit area
f	fugacity
X_i	mole fraction of ith component
r_{ij}	distance between atoms i and j
P_{in}	internal pressure
P_{ext}	external pressure
V_{in}	internal volume
V_{ext}	external volume
e_i	charge of atom
γ_{lg}	interfacial tension at liquid–gas interface
γ_{sg}	interfacial tension at solid–gas interface
γ_{sl}	interfacial tension at solid–liquid interface
k_w	wetting coefficient
$\gamma_{\ell,0}$	surface tension of liquid in vacuum
$\gamma_{\text{s},0}$	surface tension of solid in vacuum
$\mathbf{f}$	force
x	displacement
K	force constant
$\mathbf{V}$	potential energy
$\mathscr{T}$	kinetic energy
Z	partition function
p	momentum
g	multiplicity
k_B	Boltzmann constant
$\bar{E}$	average energy
$\bar{n}$	average number of quanta per oscillator
Θ_E	Einstein temperature
Θ_D	Debye temperature
ν_D	Debye characteristic frequency
ν_E	Einstein frequency

$g(\nu)$	frequency distribution
t	time
Θ_{bulk}	bulk Debye temperature
$\Theta_{surface}$	surface Debye temperature
ℓ	length
$\langle x^2 \rangle$	mean-square displacement
ν_0	atom vibrational frequency
ΔE_D^*	activation energy for surface diffusion
ΔE_f	energy of ad-atom formation
ΔE^s	relaxation energy
$\mathbf{P}$	probability
C_V	heat capacity at constant volume
δ	surface layer thickness
a_0	Bohr radius
z	number of equivalent neighboring sites
D	diffusion coefficient for surface atoms
D_0	diffusion constant
ΔG_D^*	activation free energy for surface diffusion
ΔS_D^*	activation entropy for surface diffusion
ΔG_f	free energy of ad-atom formation
Q	total activation energy
F	flux of atoms
D_V	volume diffusion coefficient
ΔH_S	enthalpy of sublimation
T_M	melting point
ψ	electron wave function
ϵ	dielectric constant
R_∞	Rydberg constant $= 13.6\,\text{eV}$
m^*	effective mass
σ	conductivity
$\mathbf{E}$	electric field
j	flux of charge carriers
v_d	drift velocity
μ	mobility
e	unit charge
l	mean free path
τ	relaxation time
n_e	electron concentration
n_h	hole concentration
μ_e	electron mobility
μ_h	hole mobility
E_c	energy of conduction band
E_v	energy of valence band

N_c	density of states of electrons
N_v	density of states of holes
E_{gap}	band-gap energy
m_e	electron effective mass
m_h	hole effective mass
I_0	incident intensity
E_F	Fermi energy
E_i	energy associated with impurity centers
N_{imp}	density of impurity centers
$\mathbf{V}_S$	surface potential
ϵ_0	permittivity of free space
l_D	Debye length
ρ	charge density
N_D^+	concentration of ionized donors
N_A^-	concentration of ionized acceptors
ϕ	work function
$\mathbf{V}_{\text{ion}}$	ionization potential
ΔH_{des}	heat of desorption
S	electron affinity
N_A	Avogadro's number
τ_0	period of single atom vibration
χ	diamagnetic susceptibility
α	polarizability
$\boldsymbol{\alpha}$	absorption coefficient
$\boldsymbol{\mu}$	dipole moment
$\bar{c}$	average velocity
$\langle \bar{u}^2 \rangle^{1/2}$	root-mean-square velocity
u	velocity of gas atom
ΔH_{ads}	heat of adsorption
μ	mass ratio
Λ	total collision rate
n_s	concentration of surface atoms
n_g	concentration of gas atoms
$F(u)$	velocity distribution of gas atoms
$G(v)$	velocity distribution of surface atoms
T	surface temperature
T_g	gas temperature
α_E	energy accommodation coefficient
$\alpha(T)$	thermal accommodation coefficient
σ	surface coverage
Θ	degree of covering
$\Delta H_{\text{ads}}^{\text{diff}}$	differential heat of adsorption
Π	surface pressure

1

Structure of

Solid Surfaces

1.1 Introduction

THROUGHOUT THE CENTURIES people have been intrigued by solid
and liquid surfaces because of their color and high degree of reflectivity.
Crystal surfaces have attracted special attention because of their growth
and ease of cleavage along certain directions at well-defined angles to a
given surface plane. To the naked eye one face of a diamond single
crystal, bounded by high-density atomic planes, looks smooth and
perfect. However, closer inspection of any freshly grown crystal surface
using an optical microscope with fairly large magnification (200×)
reveals large regions where the atoms are "dislocated" from their
position in the surface plane to other parallel atomic planes separated
from each other by ledges 10^2–10^5 Å high. For example, Fig. 1.1 shows

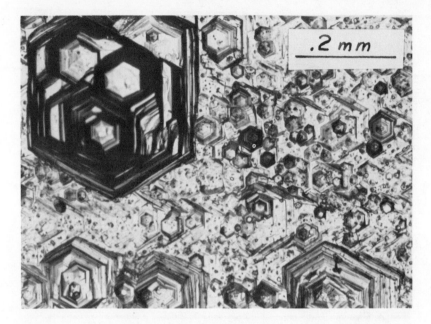

Figure 1.1. *Optical microscope picture of the basal plane of cadmium sulfide at 200 × magnification.*

one face of a cadmium sulfide single crystal that exhibits a large concentration of these terraces. The growth of these terraces of parallel atomic planes is largely due to a small mismatch of atomic planes, called *dislocations*, in the bulk of the crystal.[1] These dislocations may propagate to the surface, and growth of new atomic planes can begin at the emergence of these "line defects."

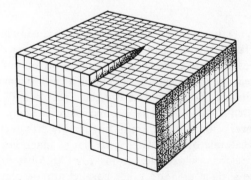

Figure 1.2. *Screw dislocation at the surface.*

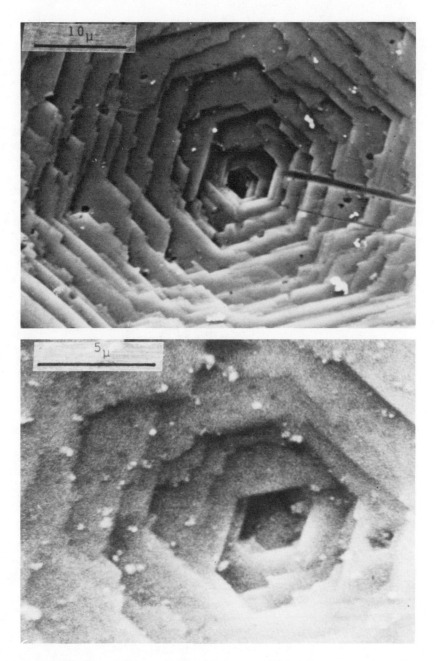

Figure 1.3. *Electron-microscope picture of the basal plane of zinc at $10^5 \times$ magnification.*

There are several types of dislocations, depending on the way the mismatch has occurred in the bulk. One type, the *screw dislocation*, is shown schematically in Fig. 1.2. Dislocation densities of the order of 10^6–10^8 cm^{-2} are commonly observable on metal or ionic single-crystal surfaces, whereas smaller dislocation densities of 10^4–10^6 cm^{-2} commonly occur in most insulator or semiconductor crystals because of their different chemical bonding properties. These concentrations may be compared with the surface concentration of atoms, which is on the order of 10^{15} cm^{-2}. Thus each terrace that has developed out of a dislocation may contain roughly $(10^{15}/10^6) = 10^9$ atoms in a low-dislocation-density single-crystal surface.

Using electron microscopy[2] to examine the structure of the seemingly

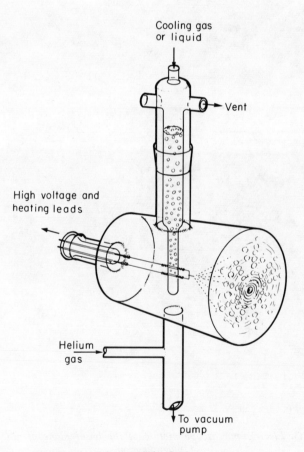

Cooling gas
or liquid

Vent

High voltage and
heating leads

Helium
gas

To vacuum
pump

Figure 1.4. *Scheme of the field-ion microscope.*

ordered surface domains between dislocations at still larger magnification, we find that these atomic terraces are far from perfect. Figure 1.3 is a scanning-electron-microscope picture of a zinc crystal plane using a magnification of about 100,000. The surface is full of ledges—small stacks of terraces separated by steps 5–100 Å high. Thus the surface is heterogeneous even at this submicroscopic scale. Typically, in an area of $1 \mu^2$ ($1 \mu = 10^{-4}$ cm) one can distinguish several surface sites which differ by the number of neighbors surrounding them. (A surface atom is surrounded by the largest number of neighbor atoms when it is located in an atomic plane; this number is reduced substantially for surface atoms along a ledge or a step.)

Now let us consider the experimental techniques capable of viewing the surface on an atomic scale. Among these the most frequently used are field-ion microscopy[3] (FIM) and low-energy electron diffraction (LEED).

Field-ion microscopy allows us to distinguish individual atoms in the different crystal planes. Under the influence of a large electric field ($\sim 10^9$ V/cm) at a small crystal tip, helium atoms that are incident on the tip are ionized. The positive ions are then repelled from the surface radially and accelerated onto a fluorescent screen, where a greatly magnified image of the crystal-tip surface is displayed. The process is shown schematically in Fig. 1.4. The intensity on any part of the screen is proportional to the number of incident helium ions. Since the concentration of helium ions produced at each crystal surface depends upon the unique atomic and electron densities in the different surface planes, the various crystal surfaces of the tip end, even the various atomic positions, can be identified by studying the intensity contrast of the fluorescent screen. (The principles of operation of field-emission and field-ion microscopy are discussed in more detail in Chapter 4.)

Figures 1.5 and 1.6 depict the image of a tungsten field-ion tip. The uniform intensity in well-defined areas of the crystal surface is due to ion emission from given atomic planes that bound the tungsten tip. The fact that these crystal faces can be readily distinguished and identified indicates that they are ordered on an atomic scale; i.e., most of the surface atoms in any given crystal face are situated in ordered rows separated by well-defined interatomic distances. The field-ion picture, however, also indicates the movement of a small number of surface atoms and the presence of point defects or vacancies in the surface. Figure 1.5(a) shows the field-ion picture of the well-ordered atomic arrangement in a tungsten tip. In Fig. 1.5(b) the same tungsten tip is shown after being disordered by helium ion bombardment. Surface atoms appear in positions that were unoccupied before ion bombardment. This can be best noticed by superimposing Fig. 1.5(a) and (b); the resultant picture is shown in Fig. 1.6. The new features that appear in the field-ion picture identify the

(a)

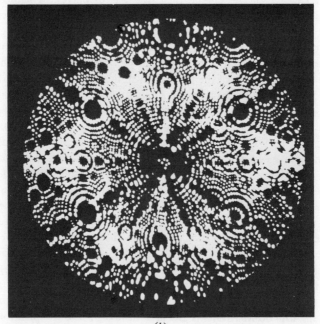

(b)

Figure 1.5. (a) *Field-ion emission pattern of a tungsten tip imaged with helium at 12°K. (b) Field-ion emission pattern of a tungsten tip after being damaged by helium ion bombardment.*

6

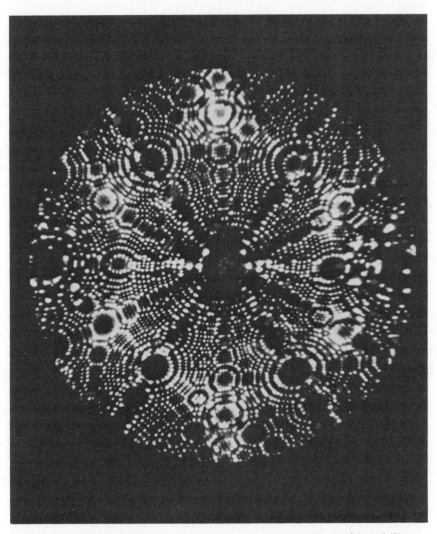

Figure 1.6. *Superposition of field-ion emission patterns in Fig. 1.5(a) and (b) to identify the new positions of surface atoms.*

new positions of the surface atoms. Surface atoms may move along one atomic plane and leave a vacancy—a vacant atomic position—behind. Thus, along with the atomic steps and ledges, single atoms and vacancies (point defects) are also distinguishable in atomically ordered crystal planes.

Another method which has revealed that the crystal surfaces are ordered on an atomic scale and which is most frequently used to study the structure of surfaces utilizes diffraction. In order to obtain diffraction from surfaces, the incident wave has to satisfy the condition $\lambda \leqslant d$, where λ is the wavelength of the incident beam and d is the interatomic distance in the surface. Also, the incident beam should not penetrate much below the surface plane but should back-diffract dominantly from the surface so that the scattered beam reflects the properties of the surface atoms and not of atoms in the bulk. Low-energy electrons satisfy these conditions and low-energy electron diffraction is used almost exclusively at present to study the structure of crystal surfaces and the rearrangement of surface atoms which may take place as a function of temperature or exposure to gases.

The technique of a low-energy electron diffraction experiment is shown in Fig. 1.7, in which monochromatic electrons are shown backscattering

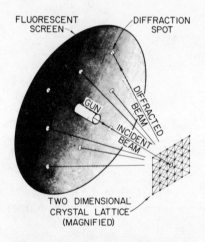

Figure 1.7. *Scheme of the low-energy electron diffraction technique.*

from one face of a single crystal. The elastically scattered fraction, i.e., those electrons that do not lose energy in the scattering process, contains the diffraction information. These electrons are allowed to impinge on a fluorescent screen on which the diffraction pattern is displayed.[4] A typical diffraction pattern from a clean (110) face of a chromium single crystal is shown in Fig. 1.8. The more that atoms are located in ordered rows in which they are separated from their neighbors by well-defined distances, the smaller are the diffraction spots and the higher their intensity. The presence of the sharp diffraction spots clearly indicates that the surface is ordered on an atomic scale. Similar low-energy electron diffraction

Figure 1.8. *Low-energy electron diffraction pattern. From a clean (110) face of a chromium single crystal at about 60eV.*

patterns have been obtained from solid single-crystal surfaces of many types.

Thus it can be concluded that

1. The surface is heterogeneous on a microscopic and submicroscopic scale. One can distinguish atomic terraces separated by ledges—atomic steps of various heights. There are vacancies in surfaces and several atomic positions distinguishable by their different numbers of nearest neighbors.

2. The surface appears to be well ordered on an atomic scale. Most of the surface atoms are located in ordered rows characterized by well-defined interatomic distances.

Later in this chapter the different ordered arrangements surface atoms can take up, as revealed by low-energy electron diffraction studies, will be discussed. (In Chapter 5 the structure of adsorbed gases on surfaces will be described.) It is essential, however, that we first familiarize ourselves with the symmetry properties and nomenclature of two-dimensional surface structures and learn about the nature of low-energy electron diffraction.

1.2. Symmetry Properties of the Two-dimensional Lattice

We are looking for a simple way to describe the two-dimensional surface structure. Since the diffraction pattern is a consequence of scattering by ordered arrangements of atoms, we shall be looking for the collection of symmetry elements that characterize a surface structure; i.e., we are looking for its space group.[5,6]

One of the symmetry operations that allows us to generate the two-dimensional structures is translation. Consider a two-dimensional array of lattice points and two translational vectors **a** and **b** as given in Fig. 1.9.

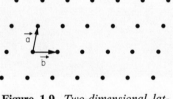

Figure 1.9. *Two-dimensional lattice with the translation vectors **a** and **b**.*

We define the translational operation **T** as

$$\mathbf{T} = n_1\mathbf{a} + n_2\mathbf{b} \tag{1.1}$$

where n_1 and n_2 are integers. If we are initially at point r at the start, the translational operation for any values of n will take us into an environment identical to that in r. By letting n_1 and n_2 vary, we can generate an infinite array; i.e., we can generate the surface structure. The parallelogram defined by the translational vectors is called the *primitive unit cell*. The translational vectors are called *primitive* if *every identical site* in the surface can be reached by an application of **T** with a suitable combination of n's.

Other symmetry operations—the point operations—do not involve translation. In two dimensions rotation about an axis and reflection through a line are the only two point operations. Inversion, which is rotation followed by reflection about a mirror plane perpendicular to the rotation axis, cannot take place in two dimensions.

A structure possesses rotational symmetry about an axis if after rotation by an angle θ the structure appears as it did prior to rotation. In order for the structure to remain invariant by rotation, θ must be some integer fraction of 2π, i.e., $\theta = 2\pi/n$, where n is an integer called the *multiplicity* of the rotation axis. It may be shown that perfect two-dimensional symmetry allows for only a finite number of different types of rotations (even though each allowed rotation may be applied an infinite number of times). Only those rotations through an angle of $2\pi/n$, where $n = 1, 2, 3,$ 4, and 6, are allowed. It is easily seen that if rotations for $n = 4$ are allowed, then the lattice must be square. Similarly, rotations through $2\pi/n$ for n equal to 3 or 6 must be associated with hexagonal lattices. Arrays of scalene triangles (onefold) and equilateral triangles (threefold) will also have twofold and sixfold rotational symmetry, respectively. These are shown in Fig. 1.10. A fivefold rotational axis cannot occur since the generated surface structure which should consist of pentagons would not fill all existing space in two dimensions, as shown in Fig. 1.11.

A structure possesses reflection symmetry about a line if it remains unchanged after being reflected through that line. In the usual point-group notation the symbol m designates the reflection symmetry.

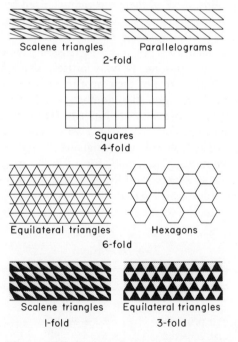

Scalene triangles   Parallelograms

2-fold

Squares

4-fold

Equilateral triangles   Hexagons

6-fold

Scalene triangles   Equilateral triangles

1-fold   3-fold

Figure 1.10. *Various lattices showing the possible one-, two-, three-, four-, and sixfold rotational axis.*

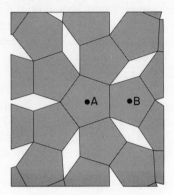

Figure 1.11. *Demonstration that a fivefold axis cannot exist since a connected array of pentagons will not fill all space.*

These different point operations yield 10 two-dimensional crystallographic point groups. Five of them are characterized by the permissible rotations, $2\pi/n$, for $n = 1, 2, 3, 4$, and 6; the other five are characterized by the permissible rotations *and* by mirror reflections.

There are only five lattice types in two dimensions; these are commonly called the five two-dimensional Bravais lattices. These lattice types are shown in Fig. 1.12 and some of their properties are also listed in Table 1.1. In point-group notation the number stands for the rotational multiplicity. For even rotational multiplicity (two-, four-, and six-fold rotations) the structure will always have zero (*m* equals zero) or two independent sets of reflection lines (designated by *mm*). Hence we have the point-group notation 2 or 2*mm*, for example. For odd rotational multiplicities (one- and three-fold) the structure contains zero or one set of reflection lines (designated by *m*).

Table 1.1. *Some Properties of the Five Bravais Lattices in Two Dimensions*

Lattice	Crystal axes		Point group
Oblique	$a \neq b$	$\phi \neq 90°$	2
Rectangular primitive	$a \neq b$	$\phi = 90°$	2*mm*
Rectangular centered	$a \neq b$	$\phi \neq 90°$	2*mm*
Square	$a = b$	$\phi = 90°$	4*mm*
Hexagonal	$a = b$	$\phi = 120°$	6*mm*

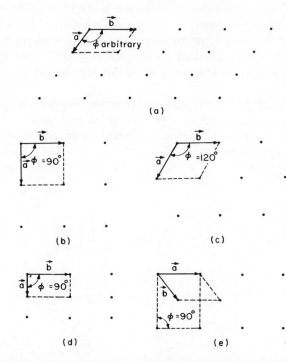

Figure 1.12. *The five two-dimensional Bravais lattices: (a) oblique (twofold rotational symmetry); (b) square (fourfold); (c) hexagonal (sixfold); (d) rectangular (twofold); and (e) centered rectangular (twofold) lattices.*

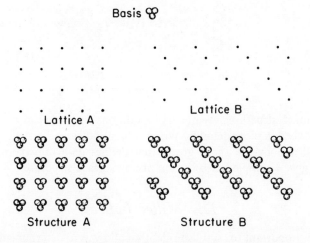

Figure 1.13. *Addition of the basis to lattices A and B to generate the surface structures A and B.*

When we locate real atoms or molecules at the idealized lattice points, the repeating unit assembly is called a *basis* (see Fig. 1.13). When we add the basis to the five Bravais lattices, the lattice symmetry may remain the same or may be reduced so that the point group of the surface structure may not be the same as that of the corresponding Bravais lattice.

The total symmetry of a crystal surface is described by the combination of the Bravais lattice and the crystallographic point group of the basis. There are 17 unique and allowed combinations of the 5 Bravais lattices and 10 crystallographic point groups. These are called two-dimensional space groups and are shown in Fig. 1.14. The crystallography of two-

Figure 1.14. *The 17 possible two-dimensional space groups.*

dimensional structures, symmetry operations, and special conventions are described in more detail by Wood.[7] Identification of a structure by its unit cell and by the space group that defines its symmetry properties should give a complete description of any surface structure.

1.3. Miller Indices

It is now important to establish rules that allow us to determine directions and the positions of atomic rows with respect to each other from the properties of the unit cell. For this purpose we use reciprocals or Miller indices; this notation has many advantages in describing diffraction and in analyzing diffraction patterns. The one important advantage that concerns us for the moment is that by using reciprocals we can describe all the atomic rows in the crystal surface in terms of lines passing through the unit cell. That is, we can generate the whole surface from the properties of atomic rows as they pass through the unit cell.

We shall first define the Miller indices in two dimensions. Let us determine the orientation of a line that runs through the crystal. First we find the intercepts of the line on the crystal axes expressed as integral multiples of the lattice constants. For the case in Fig. 1.15 we have

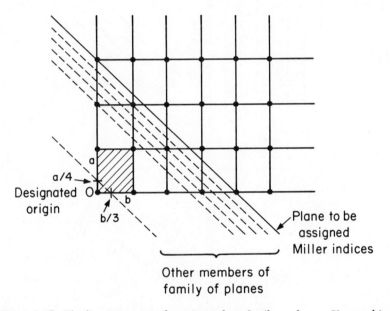

Figure 1.15. *The line intercepts the unit mesh at 3a,4b as shown. Upon taking the reciprocals of the numbers that specify the line* $(\frac{1}{3}, \frac{1}{4})$ *we reduce these fractions to the smallest two integers having the same ratio (4, 3) to obtain the Miller indices. This means that the line so defined intercepts the unit cell at a/4,b/3.*

$3a,4b$. We have now two numbers that specify the atomic row. Second, we take the reciprocals of these numbers (for the above example, $\frac{1}{3}$ and $\frac{1}{4}$), and then reduce these fractions to the smallest two integers having the same ratio. These are the Miller indices (h, k). In essence we place the reciprocals over the lowest common denominator, e.g., 12, and then discard it while retaining the numerators as the Miller indices. For our

example the Miller indices are thus 4 and 3. The line defined in such a
way intersects the unit cell at a/h and b/k. The Miller indices also define
a set of parallel lines separated from each other by the distance from the
origin of the unit cell to the nearest line:

$$d = \frac{1}{\sqrt{h^2/a^2 + k^2/b^2}} \tag{1.2}$$

as can be ascertained by inspection of Fig. 1.15. In Fig. 1.16 we give
examples of several sets of atomic rows.

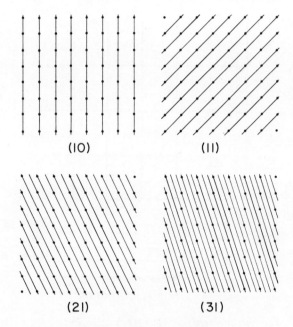

(10) (11)

(21) (31)

Figure 1.16. *Examples of several atomic rows.*

For defining crystal planes in a three-dimensional crystal as opposed to
atomic rows (or lines) on the surface we have to use three Miller indices
(h, k, l). The rules for applying them are the same as those in two
dimensions and are described in detail in several excellent books.[5,8]
Several examples of crystal planes in three dimensions are given in Fig.
1.17. If a plane is parallel to a crystal axis and thus has no intercept (the
intercept is at infinity), the Miller index for that axis is zero. For example,
the (110) family of planes are all parallel to the z axis (see Fig. 1.18).

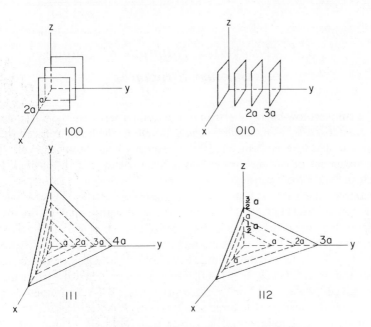

Figure 1.17. *Examples of several cubic crystal planes.*

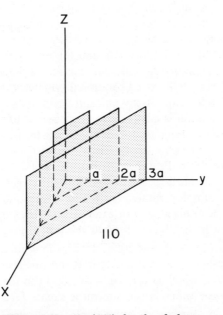

Figure 1.18. *The (110) family of planes.*

1.4 Abbreviated Notation of Simple Surface Structures

The simplest low-energy electron diffraction patterns are most frequently characterized by using a *short-hand notation* in which the unit cell of the surface structure is designated with respect to the bulk unit cell. The arrangement of surface atoms that is identical to that in the bulk unit cell is called the "substrate" structure and is designated (1×1). For example, the substrate structure of platinum on the (111) surface is designated Pt(111)-(1×1). If the surface structure that forms in the presence of an adsorbed gas is characterized by a unit cell identical to the primitive unit cell of the substrate, the surface structure is denoted (1×1)-S, where S is the chemical symbol or formula for the adsorbate [for example, Si(111)-(1×1)-O]. Other arrangements of surface atoms are called "surface net" or "surface structure." It is a common observation that the surface structures are frequently characterized by unit cells that are integral multiples of the substrate unit cell. If the unit cell of the surface structure is twice as large as the underlying bulk unit cell, it is designated (2×2). A (2×2) surface structure formed by an adsorbed gas such as hydrogen on the (211) face of tungsten is designated W(211)-(2×2)-H. If the unit cell that characterizes the surface structure is twice as long as the bulk unit cell along one major crystallographic axis and has the same length along the other, the surface structure is designated (2×1). In Fig. 1.19(a), (b), and (c), examples are shown for the most frequently occurring different surface structures on substrates having two-, four-, and sixfold rotational symmetry, respectively, where *a* is the magnitude of the X-ray unit cell vector.

This simple notation is adequate to give the size of the unit cell of the surface structure as long as it is in registry with the substrate unit cell. It is, however, not easily applicable if the surface structure is rotated with respect to the bulk unit cell or if the unit cell dimensions of the substrate and the surface net are no longer integer multiples of each other. For the simplest rotated surface structures the abbreviated notation may still be applicable. For example, if every third lattice site on a hexagonal face is distinguished from the other sites, as shown in Fig. 1.19(c), then a $(\sqrt{3} \times \sqrt{3})$-R 30° surface structure may arise. The angle given after the notation and the letter R indicate the orientation of the surface structure, which is rotated with respect to the original substrate unit cell. If every other lattice site on a square face is unique, then a $(\sqrt{2} \times \sqrt{2})$-R45° surface structure could be formed [Fig. 1.19(b)]. To

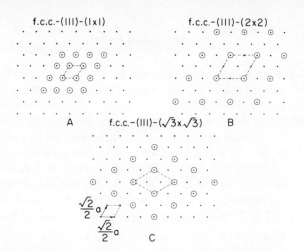

Figure 1.19a. *Schematic diagrams of surface structures on substrates with two-fold rotational symmetry.*

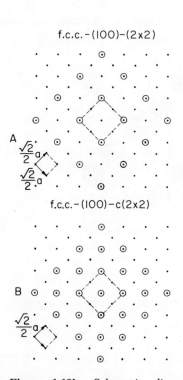

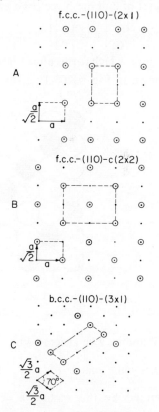

Figure 1.19b. *Schematic diagrams of surface structures on substrates with fourfold rotational symmetry.*

Figure 1.19c. *Schematic diagrams of surface structures on substrates with sixfold rotational symmetry.*

avoid noninteger notation for this structure, which occurs very frequently, it is usually labeled c(2 × 2), where the c indicates that this is a centered (2 × 2) structure [Fig. 1.19(b)].

For designating more complex surface structures a matrix notation may be used.[9] Any surface structure could be generated by one or more sets of unit mesh vectors, **a** and **b**, whose components are given by the rows of the transformation matrices. These unit mesh vectors are defined relative to the substrate unit mesh. For example, the unit cell of a (1 × 1) and a (2 × 2) surface structure on the (100) face of a face-centered cubic crystal [as in Fig. 1.19(b)] can be generated by the vectors $\mathbf{x} = 1 \cdot \mathbf{a} + 0 \cdot \mathbf{b}$, $\mathbf{y} = 0 \cdot \mathbf{a} + 1 \cdot \mathbf{b}$ and $\mathbf{x} = 2\mathbf{a} + 0 \cdot \mathbf{b}$, $\mathbf{y} = 0 \cdot \mathbf{a} + 2 \cdot \mathbf{b}$, respectively, where **a** and **b** are the unit mesh vectors that generate the substrate unit cell. Thus a (1 × 1) or a (2 × 2) surface structure can be generated by a set of unit mesh vectors **a** and **b** whose components are given by the matrices $\begin{vmatrix} 10 \\ 01 \end{vmatrix}$ and $\begin{vmatrix} 20 \\ 02 \end{vmatrix}$, respectively. A more complex rotated surface structure, which forms after heating the (100) face of platinum in oxygen, is shown in Fig. 1.20. This surface structure is denoted $(\sqrt{2} \times 2\sqrt{2})$-

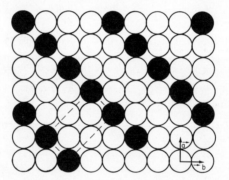

Figure 1.20. *Schematic representation of the $(\sqrt{2} \times 2\sqrt{2})$-R 45° surface structure.*

R 45° and may be generated by unit mesh vectors whose components are given by the matrix $\begin{vmatrix} -1 & 1 \\ 2 & 2 \end{vmatrix}$. In general, if the surface structure is generated by the vectors $\mathbf{a} = a_1\mathbf{x} + a_2\mathbf{y}$, $\mathbf{b} = b_1\mathbf{x} + b_2\mathbf{y}$, the transformation matrix is

$$A = \begin{vmatrix} a_1 & a_2 \\ b_1 & b_2 \end{vmatrix}$$

1.5. The Reciprocal Lattice

When a beam of incident electrons, x rays, or atoms of suitable wavelength are diffracted by an ordered surface or by a three-dimensional crystal, the diffracted beams appear as spots on a fluorescent screen or photographic plate detector. These diffraction spots appear in positions that are lattice points of a "reciprocal lattice." The properties (structure and symmetry) of this reciprocal lattice, however, are directly related to and can be obtained from the structural and symmetry properties of the real lattice from which diffraction has occurred. Thus we should proceed to examine the relationship between the real space and the reciprocal lattice structures since we could then predict the position and the symmetry of the diffraction features. In diffraction experiments, however, we proceed in just the opposite way: We obtain a diffraction pattern from a surface of unknown structure and from the properties of the diffraction features we seek to know the real-space surface structure. Thus in order to analyze the diffraction pattern, we must define the properties of the reciprocal lattice in terms of the real lattice. We give the properties of the reciprocal lattice below, without proof. For detailed analysis of reciprocal space the reader is referred to several excellent books on the subject.[10]

1. *Every reciprocal lattice vector is normal to a lattice plane of the real crystal lattice.* We define reciprocal lattice vectors for a three-dimensional reciprocal lattice as

$$\mathbf{G} = h\mathbf{A} + k\mathbf{B} + l\mathbf{C} \tag{1.3}$$

while $\mathbf{T}$, the translation vector which defines the real lattice, is given as $\mathbf{T} = n_1\mathbf{a} + n_2\mathbf{b} + n_3\mathbf{c}$.
By definition we have

$$
\begin{array}{lll}
\mathbf{A} \cdot \mathbf{a} = 2\pi & \mathbf{B} \cdot \mathbf{a} = 0 & \mathbf{C} \cdot \mathbf{a} = 0 \\
\mathbf{A} \cdot \mathbf{b} = 0 & \mathbf{B} \cdot \mathbf{b} = 2\pi & \mathbf{C} \cdot \mathbf{b} = 0 \\
\mathbf{A} \cdot \mathbf{c} = 0 & \mathbf{B} \cdot \mathbf{c} = 0 & \mathbf{C} \cdot \mathbf{c} = 2\pi
\end{array}
\tag{1.4}
$$

Thus we see that $\mathbf{A}$ must be perpendicular to $\mathbf{b}$ and $\mathbf{c}$, $\mathbf{B}$ must be perpendicular to $\mathbf{a}$ and $\mathbf{c}$, and $\mathbf{C}$ must be perpendicular to $\mathbf{a}$ and $\mathbf{b}$. Since a vector perpendicular to both $\mathbf{b}$ and $\mathbf{c}$ is given by the vector product $\mathbf{b} \times \mathbf{c}$, we have $\mathbf{A} = \alpha\mathbf{b} \times \mathbf{c}$, where α is the normalization factor. Since the condition $\mathbf{A} \cdot \mathbf{a} = 2\pi$ has to be satisfied as well, we can obtain α, $\mathbf{A}\mathbf{a} = \mathbf{a} \cdot (\mathbf{b} \times \mathbf{c}) = 2\pi$, which gives

$$\alpha = \frac{2\pi}{\mathbf{a} \cdot (\mathbf{b} \times \mathbf{c})}$$

Thus, we have

$$\mathbf{A} = 2\pi \frac{\mathbf{b} \times \mathbf{c}}{\mathbf{a} \cdot (\mathbf{b} \times \mathbf{c})} \tag{1.5}$$

We can express all the other reciprocal lattice vectors similarly in terms of the real-space coordinates. Since we have, by vector algebra, $\mathbf{a} \cdot (\mathbf{b} \times \mathbf{c}) = \mathbf{b} \cdot (\mathbf{c} \times \mathbf{a}) = \mathbf{c} \cdot (\mathbf{a} \times \mathbf{b})$, the other two reciprocal lattice vectors are given by

$$\mathbf{B} = 2\pi \frac{\mathbf{c} \times \mathbf{a}}{\mathbf{a} \cdot (\mathbf{b} \times \mathbf{c})} \quad \text{and} \quad \mathbf{C} = 2\pi \frac{\mathbf{a} \times \mathbf{b}}{\mathbf{a} \cdot (\mathbf{b} \times \mathbf{c})} \tag{1.6}$$

Using these definitions, we can readily compute reciprocal lattice vectors or other properties of reciprocal lattices. For example, a particular surface structure is characterized by the unit cell vectors $\mathbf{a} = 2\hat{x} + 3\hat{y}$, $\mathbf{b} = 4\hat{x}$, and $\mathbf{c} = \hat{z}$. Now let us suppose that $\mathbf{A}$ and $\mathbf{B}$ are in the same plane and $\mathbf{c}$ is a unit vector parallel to the z axis, perpendicular to the surface. We can now compute the reciprocal lattice vectors $\mathbf{A}$ and $\mathbf{B}$ that characterize the two-dimensional reciprocal lattice:

$$\mathbf{c} \times \mathbf{a} = -3\hat{x} + 2\hat{y}$$

$$\mathbf{b} \times \mathbf{c} = -4\hat{y}$$

$$\mathbf{a} \cdot \mathbf{b} \times \mathbf{c} = -12$$

Thus we have, by substitution,

$$\mathbf{A} = \frac{2\pi}{3}\hat{y}$$

$$\mathbf{B} = \frac{\pi}{2}\hat{x} - \frac{\pi}{3}\hat{y}$$

Figures 1.21 and 1.22 give the real lattice and the reciprocal lattice in two dimensions with the unit vectors. As one can see, any vector between reciprocal lattice points is perpendicular to an atomic row in the real crystal lattice.

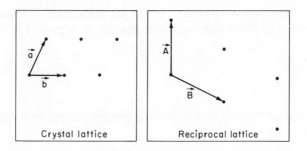

Figures 1.21 and 1.22. *Translation vectors* **a, b** *for the two-dimensional crystal lattice and* **A, B** *for the reciprocal lattice. Any vector between reciprocal lattice points is perpendicular to some lattice plane in the crystal lattice.*

By reversing this procedure the diffraction spots of a low-energy electron diffraction pattern that correspond to lattice points in the reciprocal lattice can be used to calculate the unit cell that characterizes the structure of the real surface. Let us consider the following reciprocal lattice vectors, which define the unit cell in reciprocal space:

$$\mathbf{A} = \pi\hat{x} + 2\pi\hat{y}$$

$$\mathbf{B} = -\pi\hat{x} + \pi\hat{y}$$

We shall now proceed to compute the real lattice vectors that define the real unit cell by using Eqs. (1.5) and (1.6). For two dimensions $\mathbf{C}$ is assumed to be parallel to $\hat{z}$; i.e., let $C = \pi\hat{z}$:

$$\mathbf{a} = \frac{2\pi}{\mathbf{A} \cdot (\mathbf{B} \times \mathbf{C})} (\mathbf{B} \times \mathbf{C}), \quad \text{etc.}$$

$$\mathbf{A} \cdot (\mathbf{B} \times \mathbf{C}) = 3\pi^3$$

$$\mathbf{B} \times \mathbf{C} = \pi^2(\hat{x} + \hat{y})$$

Thus

$$\mathbf{a} = \frac{2\pi}{3\pi^3} \cdot \pi^2(\hat{x} + \hat{y}) = \tfrac{2}{3}(\hat{x} + \hat{y})$$

$$\mathbf{b} = \tfrac{2}{3}(-2\hat{x} + \hat{y})$$

$$\mathbf{c} = 2\hat{z}$$

The reciprocal lattice can also be readily generated for more complex surface structures for which the transformation matrix notation is used to define the real-space surface unit mesh that generates the surface structure. In general, if the transformation matrix, A, is given by

$$A = \begin{vmatrix} a_{11} & a_{12} \\ a_{21} & a_{22} \end{vmatrix}$$

then the transformation matrix which generates the reciprocal unit cell, A^{-1}, is given by

$$A^{-1} = \frac{1}{|A|} \begin{vmatrix} a_{22} & -a_{12} \\ -a_{21} & a_{11} \end{vmatrix}$$

For example, for a (2×2) surface structure on the face-centered cubic (111) face [as shown in Fig. 1.19(c)]

$$A = \begin{vmatrix} 2 & 0 \\ 0 & 2 \end{vmatrix} \quad \text{and} \quad A^{-1} = \tfrac{1}{4} \begin{vmatrix} 2 & 0 \\ 0 & 2 \end{vmatrix}$$

Let us consider another example of a face-centered cubic (100)-c(2×2) [as shown in Fig. 1.19(b)]:

$$A = \begin{vmatrix} 1 & 1 \\ -1 & 1 \end{vmatrix} \quad \text{and} \quad A^{-1} = \tfrac{1}{2} \begin{vmatrix} 1 & -1 \\ 1 & 1 \end{vmatrix}$$

2. The spacing, d, between lattice planes (h, k, l) of the crystal is given by

$$d(hkl) = \frac{2\pi}{|\mathbf{G}(hkl)|} \tag{1.7}$$

Thus the magnitude of any reciprocal lattice vector $|\mathbf{G}(hkl)|$ is equal to $2\pi n/d(hkl)$, where n is an integer. In this way we can readily obtain the interplanar distances from the diffraction pattern.

1.6. Diffraction by Surfaces

In 1924 deBroglie suggested[11] that particles such as atoms, electrons, and neutrons might show dual character, exhibiting the properties of waves as well as those of particles. He proposed that the wavelength, λ, associated with these particles be given by

$$\lambda = \frac{h}{mv} = \frac{h}{(2mE)^{1/2}} \tag{1.8}$$

where m and v and E are the mass, velocity, and energy of the particle and h is Planck's constant. Diffraction of these particles has been observed and deBroglie's proposition has been confirmed. In Fig. 1.23, plots of the wavelength of electrons, helium atoms, x rays, and neutrons are given as functions of their energy. These values have been calculated using the deBroglie equation for the different particle waves:

$$[E(\text{eV})]^{1/2}(\text{electron}) = \frac{12}{\lambda(\text{Å})}$$

$$[E(\text{eV})]^{1/2}(\text{He atom}) = \frac{0.14}{\lambda(\text{Å})}$$

$$[E(\text{eV})]^{1/2}(\text{neutron}) = \frac{0.28}{\lambda(\text{Å})} \tag{1.9}$$

$$E(\text{eV})(\text{x ray}) = \frac{1.24 \times 10^4}{\lambda(\text{Å})}$$

The diffraction of electrons that were back-reflected from the surface of a nickel crystal was observed by Davisson and Germer,[12] diffraction of electrons transmitted through a thin metal foil was observed by G. P. Thomson and Reid,[13] and diffraction of helium atoms from LiF surfaces was first reported by Esterman and Stern.[14]

Both electrons and atoms are scattered by the atomic potential. Low-energy (5–200 eV) electrons, owing to their charge, are predominantly scattered by surface atoms. As we have mentioned previously, low-energy electron diffraction is used almost exclusively, at present, to study the structure of crystal surfaces. Helium atoms at thermal energies ($kT \approx 0.025$ eV for $T = 300°$K) are also predominantly scattered by surface atoms. Because of experimental difficulties which seem to surpass those experienced in low-energy electron diffraction studies, atomic beams are rarely used in studies to obtain information about the surface structure. However, atomic beams are most frequently used in experiments of energy transfer between the incident beam and the surface, where the energy (velocity) of the incident and scattered beams are monitored. These studies will be discussed in Chapter 5.

X rays are scattered by the electron density about the atoms. Having no charge and high energy as compared with low-energy electrons, they

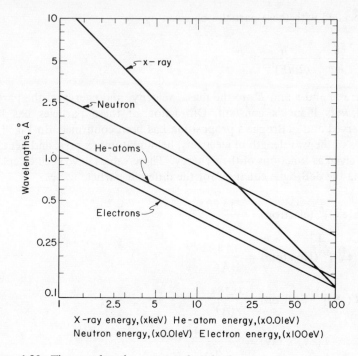

Figure 1.23. *The wavelengths associated with electrons, He atoms, neutrons, and x rays as a function of their energy.*

can penetrate deep into the bulk and show little sensitivity for surface structure. Since the atomic surface density ($\sim 10^{15}$/cm^2) is much smaller than the bulk density of atoms ($\sim 10^{22}$/cm^3) once the incident beam has penetrated over approximately 10 atomic planes, the diffracted beam contains no easily detectable information about the surface atoms. Neutrons are scattered by the atomic nuclei and are also insensitive for use in detecting the properties of the surface plane under most experimental conditions since they may also penetrate deeply into the bulk of the solid.

1.7. Low-Energy Electron Diffraction (LEED)

We shall briefly describe the conditions necessary for the diffraction of low-energy electrons from crystal surfaces. For more detailed discussions of diffraction in general and LEED in particular, the reader is referred to recent books[5,15] and reviews[4,16] on these subjects.

Consider a wave incident on a crystal surface, with an amplitude A in free space at a given point r:

$$A(r) = A(0) \exp [i(\mathbf{k} \cdot \mathbf{r} - \omega t)] \qquad (1.10)$$

Here $A(0)$ is the amplitude at an arbitrary origin, $r = 0$, and $\mathbf{k}$ is the wave vector of the traveling wave which has an angular frequency ω. The magnitude of $\mathbf{k}$ is given by $|\mathbf{k}| = 2\pi/\lambda$, where λ is the wavelength. For a given k, i.e., for incident electrons of well-defined wavelength, it can be readily shown[6,10b] that the electrons are scattered by a periodic three-dimensional lattice only in well-defined directions corresponding to wave vectors satisfying the condition

$$\mathbf{k'} = \mathbf{k} + \mathbf{G} \qquad (1.11)$$

where $\mathbf{k'}$ is the wave vector of the scattered beam. Diffraction can occur only if the magnitude of the incident and scattered wave vectors remain the same, $|\mathbf{k'}| = |\mathbf{k}|$. This is the condition of elastic scattering in which the incident electrons must not suffer any energy loss. Their momentum may change, however, as manifested by their change of direction or angle of scattering. Figure 1.24 shows typical diffraction conditions.

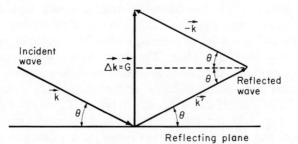

Figure 1.24. *Determination of* $\Delta \mathbf{k} = \mathbf{k'} - \mathbf{k}$ *in terms of the angle of scattering* θ.

Rearranging Eq. (1.7) we have

$$\mathbf{G}(hkl) = \frac{2\pi n}{d(hkl)} = 2 |\mathbf{k}| \sin \theta \qquad (1.12)$$

where θ is the half-angle between the scattering vectors $\mathbf{k}$ and $\mathbf{k'}$. Since $|\mathbf{k}| = 2\pi/\lambda$, we have arrived at the Bragg reflection law:

$$n\lambda = 2d \sin \theta \qquad (1.13)$$

For LEED the wave vector **k** in units of Å^{-1} is uniquely defined by the electron beam energy as

$$|\mathbf{k}| = 2\pi/\lambda = 2\pi[E(\text{eV})/150.4]^{1/2} = 0.512\sqrt{E(\text{eV})}.$$

For the case of x-ray diffraction, the scattering cross section is small ($\sim 10^{-6}$ Å^2); therefore the incident beam intensity is much larger than the intensity of the scattered beams. Thus the probability that an x ray scattered once will be rescattered is very small. As a consequence, the diffraction condition for single scattering as given by Eq. (1.11) describes the diffraction of x rays accurately. In LEED, however, where the scattering cross sections are large (~ 1 Å^2), the amplitudes of the various diffracted beams can be of the same order of magnitude as the amplitude of the incident electron beam. As a result there is a significant probability that an electron may be scattered at least twice before leaving the crystal. Thus the diffracted beams themselves may behave as "primary" beams or sources of electrons for subsequent scattering events. These double scattering events are characterized by diffraction conditions of the form

$$\mathbf{k}'' = \mathbf{k}' + \mathbf{G} \tag{1.14}$$

Here the primes refer to two different diffracted beams. Equation (1.11) is a special case of this equation. Thus it is one of the features of LEED that double diffraction or, in general, multiple scattering of low-energy electrons may also take place. The experimental verification of the importance of double diffraction and other multiple-scattering events in LEED comes from studies of curves of the intensity (I) versus electron energy (eV) of the diffracted beams. A typical I_{hk} versus eV curve is shown in Fig. 1.25 for the specularly back-reflected electron beam [(00) beam] and the (10) diffraction beam from a face-centered cubic crystal, the (100) face of aluminum.[17] [The beam for which the incident angle and the angle of scattering is identical is called the specular or (00) beam.] The appearance of strong peaks indicates that there is essentially three-dimensional diffraction. A smooth variation of the intensity with changing electron energy would be obtained for diffraction in only two dimensions. If the atomic scattering cross sections were low enough so that only single scattering events contributed to the diffraction pattern, strong peaks would be expected at the positions of the arrows, according to the three-dimensional kinematic model. There are clearly many more peaks than could be explained by this assumption. The reason the three-dimensional kinematic model is unsuccessful is that low-energy electrons are preferentially scattered from the surface layers,

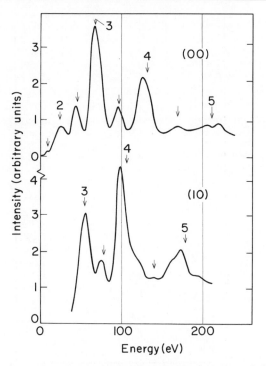

Figure 1.25. *Intensity of low-energy electrons back-scattered from the (100) crystal face of aluminum, as a function of electron energy.*

where they do not experience the full three-dimensional periodicity of the crystal potential. If the elastic scattering cross section is sufficiently high, multiple scattering events will contribute to the observed intensity. Since the elastic electron scattering cross section decreases with increasing energy,[18] the single scattering mechanism begins to dominate at higher electron energies. For other solids, such as chromium,[19] single scattering appears to dominate already above 10 eV. This is shown by the much higher intensity single scattering peaks (indicated by the arrows) as compared with the intensity of the multiple scattering peaks. This is clearly shown by the *I* versus eV curve for the (110) face of chromium in Fig. 1.26. The position of all the peaks can be predicted, assuming that both single and multiple scattering events (especially double diffraction) are taking place.

Derivation of an equation for double diffraction similar to that of the Bragg condition for single scattering is given in the Appendix to this chapter.

Low-energy electrons are produced by heating a metal or an oxide filament (cathode) in vacuum. The emitted electrons are then drawn off

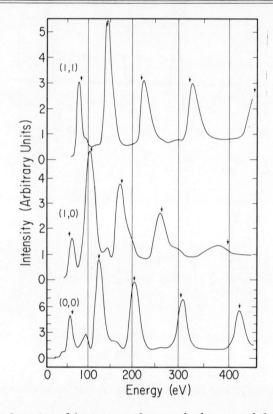

Figure 1.26. *Intensity of low-energy electrons back-scattered from the (110) crystal face of chromium, as a function of electron energy.*

toward the anode by a suitable accelerating potential. Although most of the electrons will strike the anode walls, a small fraction escapes through a hole in the anode structure. These electrons are then focused electrostatically or magnetically (approximate electron beam size, ∼1 mm²) and are allowed to impinge on the surface of a single crystal.

Figure 1.27 shows schematically the apparatus most frequently used in LEED studies. The crystal in most experiments is maintained at ground potential with respect to the negative cathode. The majority of the electrons back-scattered from the crystal have lost energy in the scattering process (inelastic scattering). These electrons contain no diffraction information and have to be separated from the elastically scattered electrons, which are back-diffracted from the solid surface. After allowing the scattered electrons to drift radially away from the crystal surface, the separation of the elastic and inelastic components is made by using a retarding potential (applied to a grid) that repels all electrons that

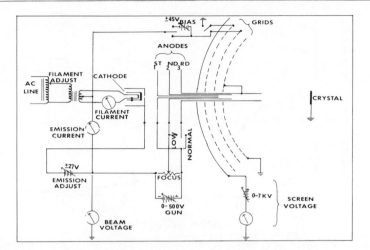

Figure 1.27. *Scheme of the low-energy electron diffraction apparatus employing the postacceleration technique.*

have lost energy in the scattering process, thus allowing only the elastically scattered electrons to penetrate. The elastically scattered electrons, after passage through the retarding field, are postaccelerated by a large positive potential and impinge on a spherical fluorescent screen where the diffraction pattern is displayed. The importance of separating the diffracted beams from the inelastically scattered fraction is indicated by Fig. 1.28, where we plot the elastic fraction of the scattered electron flux as a function of beam voltage. For a 100-eV incident beam, only about 5 per cent of the back-scattered beam is elastic, i.e., contains diffraction information. Thus, low-energy electron diffraction is taking place in the presence of a large inelastic scattering contribution.

Although the inelastic fraction of scattered electrons is discarded in a LEED experiment, these electrons are used in other experiments to detect the presence of surface impurities and measure electron binding energies in the different electron energy states for surface atoms (Auger spectroscopy). This will be discussed in Chapter 4.

Ultrahigh vacuum (pressures below 10^{-8} torr) has to be maintained in a LEED chamber to ensure that the single-crystal surface remains clean, i.e., free of adsorbed ambient gases during the time of the experiment. Assuming that every gas atom or molecule that strikes the surface adsorbs as well, the initially clean surface becomes covered by a monolayer of gas in 1 sec at an ambient pressure of 10^{-6} torr. [F (atoms/ cm² sec) $= P_{eq}(2\pi MRT)^{-1/2}$ from the kinetic theory of gases, where F is the gas flux, P_{eq} is the equilibrium pressure over the surface and M is the molecular weight of the gas in the ambient. For $P_{eq} = 10^{-6}$

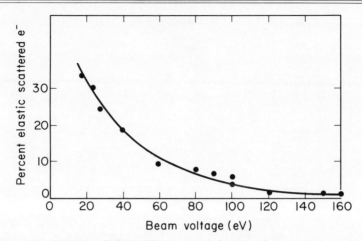

Figure 1.28. *Fraction of elastically scattered electrons from a platinum surface as a function of incident electron energy.*

torr, the flux is on the order of $10^{15}\,\text{cm}^{-2}\,\text{sec}^{-1}$ and for $P_{eq} = 10^{-9}$ torr, the flux is on the order of $10^{12}\,\text{cm}^{-2}\,\text{sec}^{-1}$.] Since typical surface densities are on the order of 10^{15} atoms/cm², the surface is covered by a monolayer of gas in 100 sec at $P_{eq} = 10^{-8}$ torr (assuming the worst case, a "sticking probability" of unity for the incident gas on the surface).

The accelerating voltage of the incident electrons can be easily varied in the range $5–10^3$ eV, thus allowing one to change the electron wavelength at will. Therefore, diffraction information can be extracted by monitoring the diffraction spot intensities as a function of wavelength. Diffraction studies may also be carried out by variation of the scattering angle at a constant electron energy.

A typical set of diffraction patterns at various electron energies from a clean (111) face of a platinum single crystal is shown in Fig. 1.29. The relative distances of the diffraction spots and their symmetry can be used to determine the size of the surface unit cell that characterizes the arrangement of surface atoms. However, this information alone is insufficient to determine the *unique* position of atoms in the unit cell, since a given diffraction pattern can be obtained from several different ordered arrangements of atoms on the surface. To determine the exact position of atoms in the unit cell, the intensities of the diffraction spots have to be analyzed based on a model calculation of the process of the low-energy electron scattering where the only important variable is the position of surface atoms. For LEED, calculations to determine the unique surface structure involve consideration of the dominant scattering characteristics: multiple scattering events, the low penetration of the electron beam (few atomic layers), and the large amount of inelastic

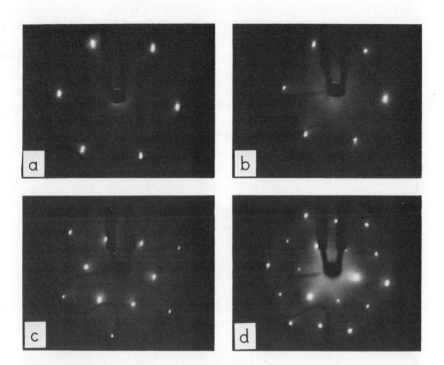

Figure 1.29. *Diffraction pattern of the (111) face of platinum single crystal at four different incident electron beam energies: (a) 51 volts; (b) 63.5 volts; (c) 160 volts; (d) 181 volts.*

scattering (Fig. 1.28). These calculations should be contrasted with computations of x-ray diffraction intensities which are carried out by considering only single scattering, deep three-dimensional penetration, and relatively small absorption corrections.

The background intensity in a LEED pattern is caused partly by the thermal-energy spread of the emitted low-energy electron beam (± 0.2 eV for a cathode temperature of 1000°K) and by the inelastically scattered electrons that suffer small energy losses (<0.2 eV) because of interactions with the vibrating atoms. The size of a diffraction spot is, to a large extent, determined by the size of the electron source. Since the source is of finite size, the lateral coherence length of electrons from such a source, ΔX (the distance within which wave pockets of electrons can undergo constructive interference), is given by[2]

$$\Delta X = \frac{2}{2\beta(1 + \Delta E/2E)} \qquad (1.15)$$

where β is the half-angle indeterminancy and ΔE is the energy spread. For $\lambda = 1$ Å, $E = 150$ eV, $\beta = 10^{-3}$ rad, $\Delta E = 0.2$ eV—which are typical values for most existing equipment—$\Delta X \approx 500$ Å. Thus the size of the diffraction spots indicates that the surface consists of ordered domains of 500 Å or larger diameter. The size of larger domains would not be detectable since the electron beam would not be coherent over a distance larger than 500 Å because of experimental limitations (size of the electron source). If the ordered surface domains are smaller than 500 Å in diameter, broadening of the diffraction spots is noticeable with a simultaneous decrease in spot intensities. Such an effect is shown in Fig. 1.30. The pattern is for the (111) face of nickel after a few minutes of

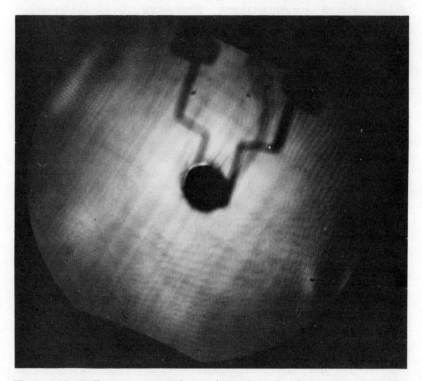

Figure 1.30. *Diffraction pattern from a (111) face of nickel single crystal after bombardment with high-energy (300 eV) argon ions at 174 eV.*

ion bombardment with high-energy argon ions (300 eV). This "ion sputtering" technique is commonly used to clean the surface inside the diffraction chamber, *in situ*.[20] Figure 1.31 was taken after heating the ion-bombarded surface in vacuum at 500°C for a few minutes. During

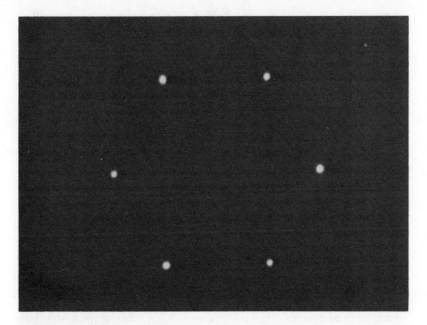

Figure 1.31. *Diffraction pattern from the (111) face of nickel crystal that has been annealed by heat treatment, after ion bombardment at 174 eV.*

the heat treatment the disordered surface anneals out; i.e., atoms in disordered positions move back to their equilibrium positions and the size of the ordered domains grows. Simultaneously, the diffraction spots sharpen up and become smaller until their limiting size, set by the use of the finite-size electron source, is obtained.

Low-energy electron diffraction from liquid metal surfaces has also been carried out. The angular distribution of the scattered electrons was virtually identical to the distribution of elastically scattered electrons from atoms in the vapor phase. Thus electron scattering due to single atoms dominates over other scattering mechanisms that are due to the presence of short-range order in the liquid surface.[21]

1.8. The Structure of Surfaces on an Atomic Scale

Most of the surfaces studied by LEED so far have been high-density, low-index crystal faces of monatomic or diatomic solids. Without exception, all these faces exhibit ordered structures on an atomic scale.

These ordered surfaces may be divided into two classes: (1) those that have unit cells identical to the projection of the bulk unit cell to the surface and (2) those characterized by unit cells that are integral multiples of the unit-cell dimensions in the bulk. Solids belonging to the first class have diffraction patterns that are characteristic of a (1×1) surface structure. The different crystal faces of tungsten [(110), (100), (211)], nickel, and aluminum [(111) and (100)], for example, seem to belong to this class. Most semiconductors and some of the metal surfaces studied so far belong to the second class. These surfaces exhibit diffraction patterns with extra diffraction features superimposed on the diffraction pattern of the substrate unit mesh (predicted by the bulk unit cell). In Table 1.2 we list some of the solid surfaces that exhibit these surface rearrangements.

Surface structures of various types on semiconductor surfaces appear to have well-defined temperature ranges of stability. At temperatures above and below this range the surface undergoes a transformation into another ordered surface structure. For example, the (111) face of silicon has a (7×7) surface structure [Si(111)-(7×7)], which forms upon heating the crystal to 700°C. The diffraction pattern corresponding to this structure is shown in Fig. 1.32. Above 800°C this surface structure transforms into a (2×2) structure; below 700°C the (1×1) surface net predominates. The (111) surfaces of semiconductors that crystallize in the diamond lattice seem to form (2×2) surface structures. What is the mechanism of surface rearrangements? There are several possible mechanisms which are under investigation; the most likely ones are discussed below.

1.8.1. SURFACE RELAXATION

Surface structures may be formed by a periodic displacement of surface atoms out of the surface plane. The surface would thus exhibit a periodic "buckling," giving rise to new, characteristic diffraction features. This mechanism may best be illustrated by considering what happens to crystal atoms in the neighborhood of a vacancy, i.e., vacant lattice position. If we remove an atom from its equilibrium position in the bulk to the gas phase, the atoms surrounding the now vacant site "relax," i.e., will be displaced slightly toward the vacancy.[22] They are no longer restrained from larger displacement in the direction of the empty site by the strong repulsive atomic potential. Therefore, the free energy of removing an atom from its bulk, equilibrium position to the gas phase is partially offset by the lattice relaxation about the vacancy. The free energy of vacancy formation from a rigid lattice that is not allowed to relax can be approximated by the cohesive energy—the energy necessary to break a solid into single atoms infinitely separated from each other.

Table 1.2. *Surface Structures of Several Semiconductor and Metal Crystal Faces*

Material	Sixfold rotation axis	Fourfold rotation axis	Twofold rotation axis
Si[16,32,33,34]	(111)-(7 × 7), (111)-(2 × 1)	(100)-(2 × 2), (100)-(4 × 4)	(110)-(5 × 2)
Ge[16,33]	(111)-(2 × 8), (111)-(2 × 1)	(100)-(2 × 2), (100)-(4 × 4)	(110)-(2 × 1)
C(diamond)[35]	(111)-(2 × 2),	(100)-(2 × 1),	
Te[36]	(0001)-(2 × 1)		
GaAs[33,37]	(111)-(2 × 2), ($\bar{1}\bar{1}\bar{1}$)-(3 × 3)		
GaSb[33,37]	(111)-(2 × 2), ($\bar{1}\bar{1}\bar{1}$)-(3 × 3)		
InSb[33,37]	(111)-(2 × 2)	(100)-(2 × 1)	
CdS[38]	(0001)-(2 × 2)		
Pt[39,40,41]		(100)-(5 × 1)	
Au[42,43,44]		(100)-(5 × 1)	(110)-(2 × 1)
Ir[45]		(100)-(5 × 1)	
Pd[44]		(100)-(1 × 1), (100)-c(2 × 2)	
Bi[46]			(1$\bar{1}$20)-(2 × 10)
Sb[46]			(1$\bar{1}$20)-(6 × 3)

(a)

(b)

Figure 1.32. *Diffraction pattern from the (111) crystal face of silicon that exhibits (a) (7 × 7) surface structure and (b) (1 × 1) surface structure.*

However, we find that the free energy of forming vacancies is always appreciably smaller than the cohesive energy for most solids where these quantities have been measured. For example, for silicon, the cohesive energy is 108 kcal/g-atom, while the free energy of vacancy formation is 53.5 kcal/g-atom. Thus the magnitude of relaxation energy can be very large.

Surface atoms are in an anisotropic environment as though they were surrounded by atoms on one side and by vacancies on the other. These atoms can relax out of plane, perpendicular to the surface—a motion that is not allowed for the bulk atoms. Depending on the bonding properties of the solid, atoms may be displaced out of plane in a periodic manner. Calculations[23,24] indicate that the formation of some of these buckled surfaces can be energetically favored over the formation of flat surfaces in temperature ranges below the melting point of the solid. One type of periodic relaxation is shown schematically in Fig. 1.33. The

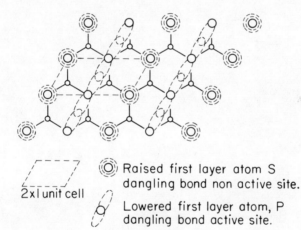

2x1 unit cell

(◎) Raised first layer atom S
dangling bond non active site.

(◌) Lowered first layer atom, P
dangling bond active site.

o Second layer atom,

Figure 1.33. *Scheme of out-of-plane surface atom relaxation that yields a (2 × 1) surface structure.*

appearance of any new surface periodicity will be reflected in the characteristics of the LEED diffraction pattern. It is likely that surface structural rearrangements in germanium, silicon, and other semiconductor surfaces take place by this mechanism.

It should be noted that the periodic out-of-plane surface relaxation should be very sensitive to the presence of impurities or to certain types of lattice defects emerging at the surface (dislocations and vacancies).

These could cause the collapse of surface structures by changing the chemical environment about the surface atoms or, in some cases, could also catalyze their formation.

1.8.2. SURFACE-PHASE TRANSFORMATION

Some of the metal surfaces were also found to undergo atomic rearrangements (see Table 1.2). For example, the (100) surfaces of gold and platinum exhibit a diffraction pattern as shown in Fig. 1.34. The presence of the $n/5$-order diffraction spots indicates the appearance of a new periodicity five times as large as that of the bulk unit cell along one principal axis in real space but the same size along the other.[25] The pattern is a result of the superposition of domains of two structures of

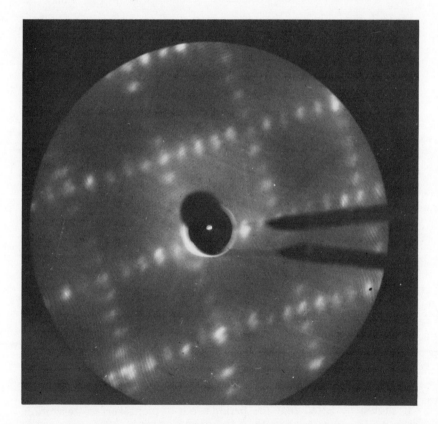

Figure 1.34. *Diffraction pattern of the (5 × 1) surface structure on the (100) crystal face of platinum at 124 eV.*

this type rotated 90° to each other. These surface structures may be designated as Au(100)-(5 × 1) and Pt(100)-(5 × 1). The diffraction patterns can be interpreted as indicating the presence of a hexagonal arrangement of scattering centers superimposed on the underlying square (100) substrate (Fig. 1.35). The interatomic spacing in the hexagonal surface layer is $\frac{5}{6}$ that of the substrate along one principal axis and the same along the other. Thus the atoms in the hexagonal surface are nearly coincident with every fifth substrate atom and this could generate the observed fivefold surface periodicity. A small compression (~5 per cent) in the hexagonal layer would allow six rows of the surface layer to fit exactly onto five rows of the square substrate. This is shown in Fig. 1.35.

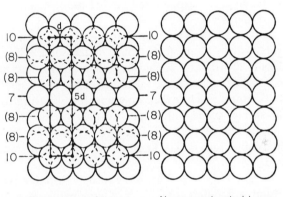

Reconstructed Layer   Nonreconstructed Layer

Figure 1.35. *Scheme of a hexagonal distortion (overlayer) of the square (100) substrate that may produce the (5 × 1) structure by coincidence of atoms in the hexagonal surface with every fifth atom in the second (substrate) layer.*

The chemical properties of this surface structure [its sensitivity to chemisorbed gases (5 × 1) $\xrightarrow[\text{absorption}]{\text{gas}}$ (1 × 1)] make it likely that the surface structure is again the result of periodic buckling of the surface plane. In this case, however, the surface relaxation resulted in the formation of a hexagonal surface structure; i.e., there is a change of rotational multiplicity (from fourfold to sixfold). For semiconductors the surface structure maintained the rotational multiplicity of the bulk unit cell, except that the surface net became enlarged. Moreover, the (5 × 1) surface structure and surface structures on other face-centered cubic metal crystals[43,45] are stable from 300°K up to the melting point of the solid. It appears that these metal surfaces have undergone a phase

41

transformation from a face-centered cubic to a hexagonal close-packed surface structure while no corresponding transformation occurs in the bulk of the solid.

The crystal structure that a solid will take up has been shown[26] to depend primarily on the number of unpaired *s* and *p* valence electrons per atom that are available for bonding. For example, solids with one (less than 1.5) unpaired *s* or *p* electrons per atom have body-centered cubic crystal structure (like Na and W). Solids with two (1.7 to 2.1) unpaired *s* and/or *p* electrons per atom will crystallize in the close-packed hexagonal structure (Zn, Os); three (2.5 to 3) unpaired valence electrons per atom will give face-centered cubic (Pt, Ag), and four unpaired valence electrons per atom give diamond crystal structures (Ge, C). A theory based on this concept, when extended to include the contribution of unpaired *d* electrons to the binding, can explain and predict the structure and stability range of most alloys.[26]

Surface atoms, in addition to being in an anisotropic environment, have fewer neighbors than atoms in the bulk of the solid. Therefore, their electron density distribution should be different from that in the bulk, and they may have fewer or more valence electrons available for bonding than the bulk atoms.[27] Thus they may undergo structural transformations in the surface plane with respect to their crystal structure in the bulk. It appears that on the (100) face of gold, platinum, and iridium a face-centered cubic close-packed hexagonal surface-phase transformation has occurred.

It should be noted again, just as in the case of surface relaxation, that impurity atoms with different numbers of unpaired valence electrons per atom may cause or accelerate surface-phase transformations of this type on transition metal surfaces, or, conversely, may inhibit it. For example, carbon (four unpaired valence electrons per atom) stabilizes the (1 $\times$ 1) surface structure on the platinum and gold (100) surfaces.[25] On the other hand, there appears to be evidence that oxygen adsorbed on these noble metal surfaces may act as an electron acceptor and can stabilize the hexagonal surface structure (5 $\times$ 1).[28]

This mechanism would also predict the formation of surface alloys with a variety of structures and other interesting physical chemical properties. These may be prepared by the deposition of other suitable metal atoms with different numbers of unpaired valence electrons.

1.8.3. STRUCTURE OF STEPPED SURFACES

When low-index crystal faces of semiconductors and metals[29] are cut at a small angle (5–15°) to a low index crystal plane, a doubling of the

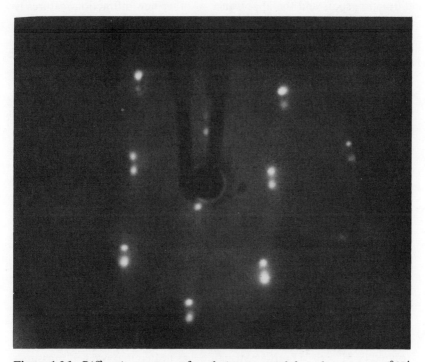

Figure 1.36. *Diffraction pattern of a platinum crystal face that was cut 6°27′ with respect to the (111) crystal face in the direction of the (110) face. Note the doubling of the diffraction spots.*

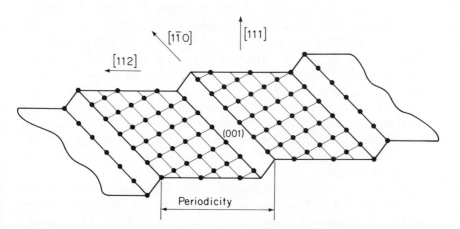

Figure 1.37. *Schematic representation of the platinum surface that exhibits ordered atomic steps.*

diffraction spots and the appearance of other new diffraction features is often observed (see Fig. 1.36). These are due to ordered arrays of atomic steps that run parallel, and their distance of separation depends on the angle of the cut with respect to the low-index surface plane. Between the steps one finds domains of the low-index surface (Fig. 1.37). These stepped surfaces show remarkable thermal stability. Since atomic steps play important roles in many surface phenomena (phase transformations, nucleation, catalysis), their study by LEED will be the subject of many investigations.

1.8.4. STRUCTURAL CHANGES DUE TO VARIATION OF SURFACE CHEMICAL COMPOSITION

There are several reports of ordered surface rearrangements induced by changes in the stoichiometry at the surface.[30,31] The study described here[31] is given as an example of this important class of surface reconstructions. The (0001) face of aluminum oxide (α-alumina), which has a sixfold rotational symmetry, exhibits a surface structure characteristic of the bulk unit cell (Fig. 1.38a). Upon heating to 1200°C in vacuum, the surface structure changes (Fig. 1.38b); during this time there is also detectable oxygen evolution from the surface. The new surface structure may be designated as $Al_2O_3(0001)$-$(\sqrt{31} \times \sqrt{31})$-R $\pm 9°$. When the rotated $(\sqrt{31} \times \sqrt{31})$ surface structure was heated in oxygen at pressures $>10^{-4}$ torr at 1100°C, the (1×1) surface structure was obtained. Removal of the oxygen and then heating to slightly higher temperature (1250°C or higher) in vacuum caused the reappearance of the rotated $(\sqrt{31} \times \sqrt{31})$ surface structure. This reversible phase transformation could be induced at will upon introduction or removal of oxygen.

When aluminum metal was condensed on the (0001) alumina surface that exhibited the (1×1) surface structure, the rotated $(\sqrt{31} \times \sqrt{31})$ surface structure formed with heating to 800°C. In the absence of excess aluminum on the surface, the (1×1) surface structure would have been stable. Thus the structural changes that occur in vacuum, in oxygen, and with aluminum indicate that the (0001) face of alumina undergoes a surface-phase transformation from a (1×1) surface structure to an oxygen-deficient, rotated $(\sqrt{31} \times \sqrt{31})$ surface structure which is stable at high temperatures.

It is difficult to explain the appearance of large surface unit cells which are also rotated with respect to the bulk unit cell without invoking significant chemical rearrangements in the surface layer. The rotated $(\sqrt{31} \times \sqrt{31})$ unit mesh signifies marked mismatch between the newly formed surface structure and the underlying hexagonal substrate.

It appears that if the high-temperature oxygen-deficient rotated

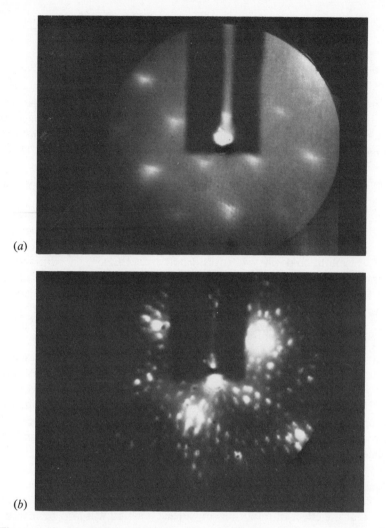

Figure 1.38. (a) *Diffraction pattern from the (0001) face of α-alumina that exhibits the (1 × 1) surface structure.* (b) *Diffraction pattern from the (0001) face of α-alumina after heating in vacuum at 1200°C, exhibiting the* $(\sqrt{31} \times \sqrt{31})$*-R* $\pm$ *9° surface.*

$(\sqrt{31} \times \sqrt{31})$ surface structure has a composition that corresponds to Al_2O (or AlO), it would be likely to form a cubic overlayer in which the cation is appreciably larger than in the underlying hexagonal (0001) substrate. Strong mismatch due to the differences in structure and ion sizes in the two layers should be expected.

If the reduced oxides of aluminum, Al_2O or AlO, are stable in the α-alumina surface at elevated temperatures, it is likely that the other thermodynamically unstable oxides might also be stable in the surface environment.

1.8.5. STUDIES OF SURFACE STRUCTURE DURING MELTING, FREEZING, AND VAPORIZATION

There are melting theories[47,48] which indicate that the surface properties are important in understanding the mechanistic aspects of the melting process. Recent kinetic studies of superheating[49,50] have shown that surfaces play an important role in initiating or nucleating melting. There is at least one model of melting[51,52] which indicates that surfaces may disorder (i.e., lose their long-range order) at temperatures below the bulk melting point.

Low-energy electron diffraction is a powerful technique to study surface melting since loss of diffraction features accompanies the melting transition. The surface melting of different low-index crystal faces of lead, bismuth, and tin was investigated by LEED.[53] The diffraction patterns that are characteristic of long-range order were detectable up to the bulk melting temperatures. Although bismuth undergoes negative volume change upon melting and has a crystal structure different from that of lead, the melting behavior of its surfaces was similar to that of lead surfaces.

The low-energy electron diffraction pattern is insensitive to the presence of disordered atoms on the surfaces as long as their concentration is only a few per cent of the total surface concentration.[4] Thus the presence of a LEED pattern from the different surfaces, which suggests the dominance of long-range order on the surface up to the bulk melting point, does not rule out the presence of disordered atoms in few atom per cent surface concentrations.

It is hoped that future studies will verify the melting behavior of high-index crystal faces that are thermodynamically less stable (see Chapter 2) than the studied low-index crystal faces of these monatomic solids and that the effect of impurities on surface melting will also be investigated.

In the same studies, the freezing rates of the molten metals were correlated with the surface structures of the growing crystallites. Slow rates ($< 0.5°C/sec$) favored the formation of the Pb(111) and Bi(01$\bar{1}$2) surfaces, while during rapid cooling ($> 0.5°C/sec$) the Pb(100) and Bi(0001) faces have predominated. The result that growth conditions far from equilibrium (fast cooling rates) produce different surface orientations of the two solids should be taken into account in future theoretical studies of crystal-growth kinetics.

Many solids sublime at an appreciable rate and have large vapor pressures below the melting point. Silver and nickel are among these solids. The surface structures of the (100) and (111) crystal faces of these two metals were monitored during vaporization by low-energy electron diffraction.[54] It was found that the vaporizing surface is ordered on an atomic scale. Although scanning-electron-microscope pictures revealed a rough surface on a 10^4-Å scale, the sharp diffraction spots in the LEED pattern indicated the presence of ordered domains of size $\geqslant 500$ Å.

Summary

Experimental evidence indicates that the crystal surface that appears smooth to the naked eye is heterogeneous on a microscopic and submicroscopic scale. There are ledges of varying heights that separate domains of atomic planes, and the presence of various point defects is also detectable. However, the crystal surface is well ordered on an atomic scale, and most of the surface atoms are located in ordered rows.

The two-dimensional lattice is characterized by a primitive unit cell. An abbreviated notation identifies the simple surface structures and a matrix notation defines the more complex surface structures with respect to the arrangement of atoms in the bulk unit cell.

Low-energy electron diffraction (LEED) appears to be most suitable to determine the atomic structure of various surfaces. The surface structures of solids that have been studied so far are either characterized (1) by unit cells identical to the projection of the bulk unit cell to the surface or (2) by a rearranged surface with a different unit cell. Surface rearrangements may take place by periodic displacement ("buckling") of atoms out of the surface plane as a result of surface relaxation or by changes in the concentration of bonding electrons per surface atom. Examples are also given of changes of surface structure that are caused by changes of surface chemical composition or by the presence of ordered steps of monatomic height.

Appendix 1

The wave vector of the incident electron beam may be expressed in a Cartesian-coordinate system as

$$\mathbf{k} = \mathbf{k}_x + \mathbf{k}_y + \mathbf{k}_z \tag{A.1}$$

Defining θ as the angle of incidence with respect to the surface normal and ϕ as the azimuthal angle, we have

$$|\mathbf{k}_x| = |\mathbf{k}| \sin \theta \sin \phi \tag{A.2a}$$

$$|\mathbf{k}_y| = |\mathbf{k}| \sin \theta \cos \phi \tag{A.2b}$$

$$|\mathbf{k}_z| = |\mathbf{k}| \cos \theta \tag{A.2c}$$

The components of wave vectors that are parallel to the surface must obey the two-dimensional diffraction grating formula:

$$\mathbf{k}'_{xy} = \mathbf{k}_{xy} + \mathbf{G}_{xy} \tag{A.3}$$

By using Eqs. (A.2a), (A.2b), and (A.3), the x and y components of the wave vector characterizing a diffracted beam can be written as

$$|\mathbf{k}'_x| = |\mathbf{k}| \sin \theta \sin \phi + |\mathbf{G}'_x| \tag{A.4a}$$

$$|\mathbf{k}'_y| = |\mathbf{k}| \sin \theta \cos \phi + |\mathbf{G}'_y| \tag{A.4b}$$

Substituting Eqs. (A.4a) and (A.4b) into the following equation, where $|\mathbf{k}_z|$ is defined,

$$|\mathbf{k}'_z| = (|\mathbf{k}|^2 - |\mathbf{k}'_x|^2 - |\mathbf{k}'_y|^2)^{1/2} \tag{A.5}$$

we obtain

$$|\mathbf{k}'_z| = \pm[|\mathbf{k}|^2 - (|\mathbf{k}| \sin \theta \sin \phi + |\mathbf{G}'_x|)^2 - (|\mathbf{k}| \sin \theta \cos \phi + |\mathbf{G}'_y|)^2]^{1/2} \tag{A.6}$$

for the component of the diffracted wave vector that is perpendicular to the surface. At normal incidence, $\theta = 0°$, Eq. (A.6) becomes

$$|\mathbf{k}'_z| = \pm(|\mathbf{k}|^2 - |\mathbf{G}'_x|^2 - |\mathbf{G}'_y|^2)^{1/2} \tag{A.7}$$

Substituting Eq. (A.7) into the diffraction equation, $\mathbf{k}'_z - \mathbf{k}''_z = \mathbf{G}$, we have

$$\pm(|\mathbf{k}|^2 - |\mathbf{G}'_x|^2 - |\mathbf{G}'_y|^2)^{1/2} \mp (|\mathbf{k}| - |\mathbf{G}''_x|^2 - |\mathbf{G}''_y|^2)^{1/2} = |\mathbf{G}_z| \tag{A.8}$$

where the appropriate signs are taken for the situation under consideration. Noting that

$$|\mathbf{k}| = 2\pi \left(\frac{eV}{150.4}\right)^{1/2}, \quad |\mathbf{G}_x| = \frac{2\pi h}{a_x}, \quad |\mathbf{G}_y| = \frac{2\pi k}{a_y}, \quad |\mathbf{G}_z| = \frac{2\pi n_z}{a_z}$$

Eq. (A.8) may be written

$$\pm\left[\frac{eV}{150.4} - \left(\frac{h'}{a_x}\right)^2 - \left(\frac{k'}{a_y}\right)^2\right]^{1/2} \mp \left[\frac{eV}{150.4} - \left(\frac{h''}{a_x}\right)^2 - \left(\frac{k''}{a_y}\right)^2\right]^{1/2} = \frac{n_z}{a_z} \tag{A.9}$$

where a_z is the distance between planes that are parallel to the surface. For the (100) face of face-centered cubic materials, $a_x = a_y = \frac{1}{2}\sqrt{2}\, a_0$ and $a = \frac{1}{2}a_0$, where a_0 is the characteristic dimension of the x-ray unit cell (e.g., 4.04 Å for aluminum). Therefore, for the metals reported here, Eq. (A.9) becomes

$$\pm\left\{\frac{eV}{150.4} - \frac{2}{a_0^2}[(h')^2 + (k')^2]\right\}^{1/2} \mp \left\{\frac{eV}{150.4} - \frac{2}{a_0^2}[(h'')^2 + (k'')^2]\right\}^{1/2} = \frac{2n_z}{a_0} \tag{A.10}$$

This equation may be solved analytically, numerically, or graphically to determine at what electron energy (in eV) the n_z diffraction condition between the (h', k') beam and the (h'', k'') beam is met.

References

1. W. T. Read: *Dislocations in Crystals*. McGraw-Hill Book Company, New York, 1953.
2. R. D. Heidenreich: *Fundamentals of Transmission Electron Microscopy*. John Wiley & Sons, Inc. (Interscience Division) New York, 1964.
3. E. W. Müller and T. T. Tsong: *Field Ion Microscopy*. American Elsevier Publishing Company, Inc., New York, 1969.
4. G. A. Somorjai and H. H. Farrell: Low Energy Electron Diffraction. *Advan. Chem. Phys.*, **20**, 215 (1970).
5. M. J. Buerger: *Crystal Structure Analysis*. John Wiley & Sons, Inc., New York, 1960.
6. R. A. Levy: *Principles of Solid State Physics*. Academic Press, Inc., New York, 1968.
7. E. A. Wood: *J. Appl. Phys.*, **35**, 1306 (1964).
8. B. D. Cullity: *Elements of X-Ray Diffraction*. Addison-Wesley Publishing Company, Inc., Reading, Mass., 1956.
9. R. L. Park and H. H. Madden, Jr.: *Surface Sci.*, **11**, 188 (1968).
10a. M. J. Buerger: *Vector Space*. John Wiley & Sons, Inc., New York, 1959.
10b. C. Kittel: *Introduction to Solid State Physics*. John Wiley & Sons, Inc., New York, 1956.
11. L. deBroglie: *Phil. Mag.*, **47**, 446 (1924).
12. C. J. Davisson and L. Germer: *Phys. Rev.*, **30**, 705 (1927); *Nature* **119**, 558 (1927).
13. G. P. Thomson and A. Reid: *Nature*, **119**, 890 (1927).
14. I. Esterman and O. Stern: *Z. Physik*, **61**, 95 (1930).
15. R. W. James: *The Optical Principles of the Diffraction of X-Rays*. G. Bell & Sons Ltd., London, 1965.
16. J. J. Lander: *Progress in Solid State Chemistry*, Vol. 2. The Macmillan Company, New York, 1965.
17a. H. H. Farrell and G. A. Somorjai: *Phys. Rev.*, **182**, 751 (1969).
17b. R. M. Goodman, H. H. Farrell, and G. A. Somorjai: *J. Chem. Phys.*, **49**, 692 (1968).
18. N. F. Mott and H. S. W. Massey: *The Theory of Atomic Collisions*. Oxford University Press, New York, 1965.
19. R. Kaplan and G. A. Somorjai: *Solid State Comm.*, **9**, 505 (1971).
20. H. E. Farnsworth: in *The Solid-Gas Interface*, E. A. Flood, ed. Marcel Dekker, Inc., New York, 1966.

21. R. M. Goodman and G. A. Somorjai: *J. Chem. Phys.*, **52,** 6331 (1970); **52,** 6325 (1970).

22. R. A. Swalin: *Thermodynamics of Solids.* John Wiley & Sons, Inc., New York, 1962.

23. D. Haneman and E. L. Heron: in *The Structure and Chemistry of Solid Surfaces*, G. A. Somorjai, ed. John Wiley & Sons, Inc., New York, 1969.

24. J. J. Burton and G. Jura: in *The Structure and Chemistry of Solid Surfaces*, G. A. Somorjai, ed. John Wiley & Sons, Inc., New York, 1969.

25. A. E. Morgan and G. A. Somorjai: *J. Chem. Phys.*, **51,** 3309 (1969).

26a. L. Brewer: *Science*, **161,** 115 (1968).

26b. L. Brewer: in *Electronic Structure and Alloy Chemistry*, P. A. Beck, ed. John Wiley & Sons, Inc. (Interscience Division), New York, 1963.

27. P. W. Palmberg and T. N. Rhodin: *J. Appl. Phys.*, **39,** 2425 (1968).

28. A. E. Morgan and G. A. Somorjai: *Surface Sci.*, **12,** 405 (1968).

29a. M. Henzler: *Surface Sci.*, **19,** 159 (1970).

29b. G. E. Rhead and J. Perdereau: *Comp. Rend.*, Ser. C, 1183 (1969).

30. L. Fiermans and J. Vennik: *Surface Sci.*, **9,** 187 (1968).

31. T. M. French and G. A. Somorjai: *J. Phys. Chem.*, **74,** 2489 (1970).

32. R. E. Schlier and H. E. Farnsworth: *J. Chem. Phys.*, **30,** 917 (1959).

33a. J. W. May: *Ind. Eng. Chem.*, **57,** 19 (1965).

33b. D. L. Heron and D. Haneman: *Surface Sci.*, **21,** 12 (1970).

34. J. J. Lander and J. Morrison: *J. Chem. Phys.*, **33,** 729 (1962).

35a. J. B. Marsh and H. E. Farnsworth: *Surface Sci.*, **1,** 3 (1964).

35b. J. J. Lander and J. Morrison: *Surface Sci.*, **4,** 241 (1966).

36. S. Andersson, I. Marklund, and D. Andersson: in *The Structure and Chemistry of Solid Surfaces*, G. A. Somorjai, ed. John Wiley & Sons, Inc., New York, 1969.

37. A. U. MacRae and G. W. Gobeli: *J. Appl. Phys.*, **35,** 1629 (1964).

38. B. D. Campbell, G. A. Haque, and H. F. Farnsworth: in *The Structure and Chemistry of Solid Surfaces*, G. A. Somorjai, ed. John Wiley & Sons, Inc., New York, 1969.

39. H. B., Lyon Jr., and G. A. Somorjai: *J. Chem. Phys.*, **46,** 2539 (1967).

40. S. Hagstrom, H. B. Lyon, Jr., and G. A. Somorjai: *Phys. Rev. Letters,* **15,** 491 (1965).

41. A. E. Morgan and G. A. Somorjai: *Surface Sci.*, **12,** 405 (1968).

42. D. G. Fedak and N. A. Gjostein: *Acta Met.*, **15,** 827 (1967).

43. P. W. Palmberg and T. N. Rhodin: *Phys. Rev.*, **161,** 586 (1967).

44. A. M. Mattera, R. M. Goodman, and G. A. Somorjai: *Surface Sci.*, **7,** 26 (1967).

45. J. T. Grant: *Surface Sci.*, **18,** 228 (1969).

46. F. Jona: *Surface Sci.*, **8,** 57 (1967).

47. I. N. Stranski, W. Gans, and H. Rau: *Ber. Bunsenges. Physik. Chem.* **67**, 965 (1963).

48. J. P. Stark: *Acta Met.*, **13**, 1181 (1965).

49. R. L. Cormia, J. D. MacKenzie, and D. Turnbull: *J. Appl. Phys.*, **34**, 2239 (1963).

50. M. Käss and S. Magun: *Z. Krist.*, **116**, 354 (1961).

51. F. A. Lindemann: *Physik. Z.*, **14**, 609 (1910).

52. J. J. Gilvarry: *Phys. Rev.*, **102**, 308 (1956).

53. R. M. Goodman and G. A. Somorjai: *J. Chem. Phys.*, **52**, 6325 (1970).

54. G. A. Somorjai: *J. Phys.*, **31**, C1-139 (1970).

Problems

1.1. Copper has a face-centered cubic crystal structure and an interatomic distance of 2.55 Å. Calculate the surface concentration of atoms (atoms/cm^2) in the (111), (100), and (110) crystal faces.

1.2. Find the Miller indices of atomic rows that intercept the crystal axes at the following distances: $2a,4b$; $1a,5b$; $5a,1b$; and $3a,7b$. Calculate the interrow distances between rows parallel to each of the above and identify the set of rows with the largest interrow spacing.

1.3. Compute the electron energies at which the (10), (11), and (21) diffraction beams from the (100) faces of aluminum are to appear for $n_z = 1$ and 2. Assume normal electron beam incidence.

1.4. Calculate the energies of electrons, helium atoms, hydrogen molecules, and x rays which correspond to a wavelength of 1.5 Å.

1.5. An exposure of gas molecules of 10^{-6} torr sec is called a *Langmuir*. How much time is necessary to cover a Pt(111) crystal face with a monolayer of carbon monoxide at pressures of 10^{-7} and 10^{-10} torr assuming that every molecule that collides with the surface sticks to it and that the surface temperature is 300°K. Calculate the exposure in each case in Langmuirs.

1.6. It has been suggested[25,28] that the (100) face of gold and platinum undergoes a face-centered cubic–hexagonal close-packed surface-phase transformation. Suggest five condensible or gaseous impurities that can aid such structural rearrangement and five impurities that can impede the transformation.

1.7. Semiconductor surfaces exhibit several ordered surface structures that are stable in vacuum in well-defined temperature ranges. It has been reported [*J. Appl. Phys.*, **40**, 3758 (1970)] that the adsorption of oxygen converts the Ge(III)-(2 × 1) surface structure to the bulk-like Ge(111)-(1 × 1) surface structure. There are reports, however, that some of the surface structures appear only in the presence of trace impurities [*Surface Sci.*, **15**, 169 (1969)]. Review and discuss those recent papers that put forward the various suggestions concerning the formation of semiconductor surface structures [*Surface Sci.*, **21**, 12 (1970); **11**, 153 (1968); etc.].

2

Thermodynamics

of Surfaces

2.1. Definition of Surface Thermodynamic Functions

ATOMS IN A SURFACE are in an environment markedly different from the environment of atoms in the bulk of the solid. They are surrounded by fewer neighbors than bulk atoms and there is an anisotropic distribution of these neighbors, which is characteristic only of a surface. We would like to define the thermodynamic properties associated with this "surface phase" separately from the bulk thermodynamic properties.

Consider a large homogeneous crystalline body that contains N atoms and is surrounded by plane surfaces. The energy and the entropy of the solid *per atom* are denoted by E^0 and S^0. The specific surface energy, E^s (energy per unit area) is defined by the relation[1]

$$E = NE^0 + \mathscr{A}E^s \tag{2.1}$$

where E is the total energy of the body and $\mathscr{A}$ the surface area.[2] Thus E^s is the excess of the total energy E that the solid has over the value NE^0, which is the value it would have if the surface were in the same thermodynamic state as the homogeneous interior. Similarly, we can write the total entropy of the solid as

$$S = NS^0 + \mathscr{A}S^s \tag{2.2}$$

where S^s is the specific surface entropy (entropy per unit area of surface created). The surface work content, A^s (energy per unit area), is defined by the equation

$$A^s = E^s - TS^s \tag{2.3}$$

and the surface free energy (energy per unit area) is defined by the equation

$$G^s = H^s - TS^s \tag{2.4}$$

where H^s is the specific surface enthalpy, i.e., the heat absorbed by the system per unit surface area created. The total free energy of a system, G, can also be expressed as

$$G = NG^0 + \mathscr{A}G^s \tag{2.5}$$

similar to the total energy and entropy in Eqs. (2.1) and (2.2). We have thus defined the thermodynamic properties of the surface as excesses of the bulk thermodynamic properties that are due to the presence of the surface surrounding the condensed phase.

2.2. *Surface Work in a One-Component System*

To create a surface we have to do work on the system that involves breaking bonds and removing neighboring atoms. Under conditions of equilibrium at constant temperature and constant pressure, the reversible surface work† δW^s required to increase the surface area $\mathscr{A}$ by an amount

---

† We use the notation δW to indicate that the work W, unlike the free energy G or other thermodynamic functions, is not independent of the reaction path, i.e., it is not a total differential.

$d\mathscr{A}$, of a one-component system, is given by

$$\delta W^s_{T,P} = \gamma \, d\mathscr{A} \tag{2.6}$$

Equation (2.6) can be compared with the reversible work to increase the volume of a one-component system at constant pressure, $P \, dV$. Here γ is the two-dimensional analogue of the pressure and is called the "surface tension," while the volume change is replaced by the change in surface area. We may consider γ as a pressure along the surface plane that opposes the creation of more surface. The pressure, P, is force per unit area (dynes/cm²); therefore the "surface pressure," γ, is given by force per unit length (dynes/cm). (The customary units of surface tension, dynes/cm or ergs/cm², are dimensionally identical.) In the absence of any irreversible process the reversible work, $\delta W^s_{T,P}$, is equal to the change in the total free energy of the surface. The total surface free energy is thus equal to the specific free energy times the surface area,

$$\delta W^s_{T,P} = d(G^s \mathscr{A}) \tag{2.7}$$

Creation of a stable interface always has a positive free energy of formation. This reluctance of the solid or liquid to form a surface defines many of the interfacial properties of the condensed phases. For example, liquids tend to minimize their surface area by assuming a spherical shape. Solids, when they are near equilibrium with their own liquid or vapor, will form surfaces of lowest free energy at the expense of other surfaces of higher free energy. Crystal faces, which exhibit the closest packing of atoms, tend to be surfaces of lowest free energy of formation and hence the most stable.

In principle there are two ways of forming a new surface: (1) by simply increasing the surface area or (2) by stretching the already existing surface (as if it were a rubber mat) with the number of atoms per unit area fixed and thereby altering the state of strain.[3] We can rewrite the right-hand side of Eq. (2.7) to give

$$\delta W^s_{T,P} = \left[\frac{\partial(G^s \mathscr{A})}{\partial \mathscr{A}} \right]_{T,P} d\mathscr{A} = \left[G^s + \mathscr{A} \left(\frac{\partial G^s}{\partial \mathscr{A}} \right)_{T,P} \right] d\mathscr{A} \tag{2.8}$$

If we create the new surface by increasing the area, the specific surface free energy G^s is independent of the surface area $(\partial G^s / \partial \mathscr{A})_{T,P} = 0$, and the surface work is given by

$$\delta W^s_{T,P} = G^s \, d\mathscr{A} \tag{2.9}$$

since the second term on the right-hand side of Eq. (2.8) vanishes. This happens when we increase the surface area by cleavage or if the solid is strained at *high temperatures*, under conditions such that atoms can diffuse to the surface and become part of the new surface. In this circumstance surface strain cannot be maintained.

Combining Eqs. (2.6) and (2.8), we have

$$\gamma = G^s \qquad\qquad (2.10)$$

that is, the *surface tension* is equal to the specific surface free energy for a one-component system. These terms are frequently used interchangeably in the literature. For solids at low temperatures, however, "cold working" of the material can lead to the formation of new surface by strain that is not relieved because of the negligible atomic diffusion rates. Thus the surface work is given by Eq. (2.8). Combining Eqs. (2.6) and (2.8) we have

$$\gamma = G^s + \mathscr{A}\left(\frac{\partial G^s}{\partial \mathscr{A}}\right)_{T,P} \qquad\qquad (2.11)$$

For one-component systems where surface strain can be produced and maintained, the surface tension is not equal to the specific surface free energy but is given by Eq. (2.11). Theoretical estimates for ionic crystals indicate[3] that $\gamma - G^s$ may often be on the order of 100 ergs/cm².

The surface tension γ, for an unstrained phase, is also equal to the increase of the total free energy of the system per unit increase of the surface area:

$$\gamma = G^s = \left(\frac{\partial G}{\partial \mathscr{A}}\right)_{T,P} \qquad\qquad (2.12)$$

The change of the total free energy, dG, of a one-component system can then be written† with the inclusion of the increase of free energy with

---

† The free energy of a one-component system is defined by the equation[1]

$$G = H - TS \qquad\qquad (a)$$

Thus the free-energy change is given by

$$dG = dH - T\,dS - S\,dT \qquad\qquad (b)$$

The enthalpy is defined as

$$H = E + PV \qquad\qquad (c)$$

increasing surface area as

$$dG = -S\,dT + V\,dP + \gamma\,d\mathscr{A} \tag{2.13}$$

A rough estimate of the surface tension can be made by assuming that the surface work is of the same magnitude as the heat of sublimation since sublimation continually creates new surface. For many metals this is in the range of 10^5 cal/g-atom. Using the energy-conversion table one obtains $10^5 \times 6.94 \times 10^{-17} = 6.94 \times 10^{-12}$ erg/atom. For a surface concentration of 10^{15} atoms/cm^2 the estimated surface tension is $6.94 \times 10^{-12} \times 10^{15} \approx 7000$ ergs/cm^2. This value may be compared with the experimental values of the surface tension of several metals that are listed in Table 2.1 along with the surface tensions of other liquids and solids that were measured in equilibrium with their own vapor. Although the estimate gives the right order of magnitude, it is too high since it does not take into consideration the "relaxation" of surface atoms in the freshly created surface. Calculations that give a more accurate estimate are discussed in Section 2.5. There are a great number of experimental methods to measure the surface tension of liquids and solids in a variety of experimental conditions (ambient, temperature, etc.). Detailed accounts of the use of these techniques can be found in the literature.[4,5] Here we only mention a few of the more frequently employed methods of surface-tension measurement.

1. The static drop method. Drops of liquid tend to become spherical because their surface tension acts to minimize surface area. In turn, suitable analysis of drop geometry allows the determination of their surface tension.

2. The ring method. The force required to detach a ring of wire from the surface of a liquid is measured, from which the surface tension is computed.

---

and therefore the change in enthalpy is given by

$$dH = dE + P\,dV + V\,dP \tag{d}$$

The reversible change in energy, dE, due to heat adsorbed by the system and expansion work by the system, is given by

$$dE = T\,dS - P\,dV \tag{e}$$

Substitution of Eqs. (d) and (e) into Eq. (b) yields

$$dG = -S\,dT + V\,dP \tag{f}$$

Table 2.1. *Surface Tension of Selected Solids and Liquids*

Material	γ (ergs/cm^2)	T (°C)
W (solid)[22]	2900	1727
Nb (solid)[22]	2100	2250
Au (solid)[22]	1410	1027
Ag (solid)[22]	1140	907
Ag (liquid)[4]	879	1100
Fe (solid)[22]	2150	1400
Fe (liquid)[4]	1880	1535
Pt (solid)[22]	2340	1311
Cu (solid)[22]	1670	1047
Cu (liquid)[4]	1300	1535
Ni(solid)[22]	1850	1250
Hg (liquid)[4]	487	16.5
LiF (solid)[23]	340	−195
NaCl (solid)[23]	227	25
KCl (solid)[23]	110	25
MgO (solid)[23]	1200	25
CaF$_2$ (solid)[23]	450	−195
BaF$_2$ (solid)[23]	280	−195
He (liquid)[4]	0.308	−270.5
N$_2$ (liquid)[4]	9.71	−195
Ethanol (liquid)[4]	22.75	20
Water[4]	72.75	20
Benzene[4]	28.88	20
n-Octane[4]	21.80	20
Carbon tetrachloride[4]	26.95	20
Bromine[4]	41.5	20
Acetic acid[4]	27.8	20
Benzaldehyde[4]	15.5	20
Nitrobenzene[4]	25.2	20

3. The drop-weight method. The volume of a drop is determined at the time of its detachment from a capillary of known orifice. Its weight and surface tension may then be calculated.

4. Capillary rise. The capillary rise of different liquids between two parallel plates or in a capillary tube is measured, and from the data their surface tensions can be determined.

5. Crystal cleavage. The isothermal work needed to increase the surface area is measured by detecting the energy necessary to cleave the crystal or to propagate a crack in a solid.

6. Calorimetric technique. When finely divided powders of large surface/volume ratio are dissolved in suitable solvents, the heats of

solution [ΔH(solution)] are decreased by the extra enthalpy associated with the surface with respect to the heat of solution of large particles having negligible surface area. From the data the specific surface energy may be calculated.

2.3. Temperature Dependence of the Specific Surface Free Energy

Equation (2.10) holds for most systems in which surface-tension measurements can conveniently be carried out and for the temperature ranges used in these studies. Differentiating Eq. (2.4) as a function of temperature, we can write

$$\left(\frac{\partial G^s}{\partial T}\right)_P = \left(\frac{\partial \gamma}{\partial T}\right)_P = -S^s \tag{2.14}$$

That is, from the temperature dependence of the surface tension we can obtain the specific surface entropy.

The surface tension of most liquids decreases with increasing temperature. This indicates that the work necessary to create more surface decreases with increasing temperature. Figure 2.1 shows a typical

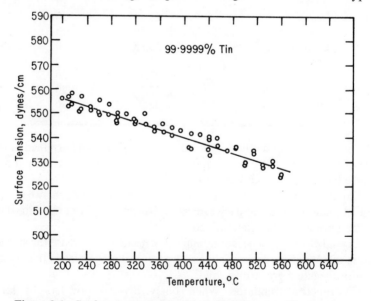

Figure 2.1. *Surface tension of liquid tin as a function of temperature.*

experimental curve for liquid tin.[6] The γ versus T curve has a negative slope that gives the specific surface entropy directly. A semiempirical equation for predicting the temperature dependence of the surface tension was proposed by van der Waals and Guggenheim[7]:

$$\gamma = \gamma^0(1 - T/T_c)^n \tag{2.15}$$

where T_c is the critical temperature and $\gamma^0 = \gamma$ at $T = 0°$K. According to Eq. (2.15), the surface tension should vanish at $T = T_c$, as expected since the interface vanishes at the critical temperature. The exponent n is determined by experiments to be near unity for metals and somewhat larger than unity for many organic liquids.[8]

There may be systems where the surface becomes more ordered with increasing temperature; such surface ordering would be indicated by the γ versus T curves. Positive slopes of the γ versus T curves have been reported from studies of the temperature dependence of the surface tension of copper and zinc.[9] On the other hand, the surface may become disordered faster than the corresponding bulk phase. In that case γ would approach zero faster than predicted by Eq. (2.15). Thus determination of γ as a function of temperature provides a great deal of information about ordering at the surface.

Substitution of Eq. (2.14) into (2.4) with subsequent rearrangement yields the specific surface enthalpy

$$H^s = G^s + TS^s = \gamma - T\left(\frac{\partial \gamma}{\partial T}\right)_P \tag{2.16}$$

Thus at constant pressure the heat adsorbed upon the creation of a unit surface area is given by Eq. (2.16). If there is no volume change associated with this process, H^s equals E^s, and the specific surface energy is given by the same equation:

$$E^s_{P,V} = \gamma - T\left(\frac{\partial \gamma}{\partial T}\right)_{P,V} \tag{2.16a}$$

Since the surface tension in general decreases with increasing temperature, the derivative $(\partial \gamma/\partial T)_P$ is negative. Therefore, the specific surface energy is somewhat larger than the specific surface free energy G^s (or γ). In theoretical calculations of surface thermodynamic properties, the specific surface energy is given more frequently than the specific surface free energy. However, G^s is determined more readily by experiments.

The temperature derivative of the specific surface enthalpy is the specific surface heat capacity:

$$C_P^s = \left(\frac{\partial H^s}{\partial T}\right)_P = T\left(\frac{\partial S^s}{\partial T}\right)_P = -T\left(\frac{\partial^2 \gamma}{\partial T^2}\right)_P \tag{2.17}$$

This equation is obtained by differentiation of Eq. (2.16) and substitution of Eq. (2.14). We can see that the specific surface heat capacity can also be expressed in terms of the temperature derivative of the surface tension. Thus accurate surface tension measurements as a function of temperature should be good sources of surface-heat-capacity data. Although many of the γ versus T curves show marked curvature, $(\partial S^s/\partial T \neq 0)$, instead of a straight line (i.e., $\partial S^s/\partial T = 0$), the data are not accurate enough to permit computation of reliable surface-heat-capacity values. Such data are more readily available from direct surface-heat-capacity measurements on finely divided powders of large surface/volume ratio.[10] When the heat capacity of the powder is compared with the heat capacity of large crystallites of the same material, the difference yields the surface heat capacity. Heat-capacity measurements on powdered samples yield larger heat capacities than measurements on samples with small surface area, as expected. However, quantitative determination of C_P^s is difficult, owing to the uncertainties of surface area measurements and the difficulties of assessing the role of strain in the surface, which could also affect the surface-heat-capacity values obtained in this manner [see Eq. (2.11)].

At low temperatures, surface heat capacity should have a different temperature dependence than the bulk heat capacity. The derivation and discussion of the temperature dependence of C_P^s will be given in Chapter 3.

2.4. Surface Tension of Multicomponent Systems

The chemical potential of the ith component, μ_i, of a multicomponent system is defined by

$$\mu_i = \bar{G}_i = \left(\frac{\partial G}{\partial n_i}\right)_{n_j, P, T} \tag{2.18}$$

where $\bar{G}_i$ is the partial molal free energy and $\partial G/\partial n_i$ is the change of the total free energy of the system with respect to changes in the number of moles of the ith component. The number of moles of all the other components, n_j, remains constant, except n_i.

The total free-energy change of a multicomponent system can be expressed, with the inclusion of the surface term, as

$$dG = \left(\frac{\partial G}{\partial T}\right) dT + \left(\frac{\partial G}{\partial P}\right) dP + \left(\frac{\partial G}{\partial \mathscr{A}}\right) d\mathscr{A} + \sum_i \left(\frac{\partial G}{\partial n_i}\right) dn_i$$

$$= -S\, dT + V\, dP + \gamma\, d\mathscr{A} + \sum_i \mu_i\, dn_i \tag{2.19}$$

Consider now the free-energy change for the process in which the surface area is increased by $d\mathscr{A}$ by transferring dn_i moles from the bulk to the surface at constant temperature and pressure. We have

$$dG_{T,P} = \gamma\, d\mathscr{A} - \sum_i \mu_i\, dn_i \tag{2.20}$$

where the minus sign indicates the decrease of the bulk concentration of the ith component. This equation shows that the surface tension γ *cannot* be identified with the total surface free energy per unit area for a multicomponent system. Equation (2.12) holds only for single-component systems or for systems in which any change in the surface area does not cause changes in the surface composition. If the bulk concentration decreases by transferring dn_i moles to the freshly created surface, there is a simultaneous increase in the number of moles of the ith component in the surface by the same amount:

$$-dn_i = c_i^s\, d\mathscr{A} \tag{2.21}$$

where c_i^s is the surface concentration in units of moles per square centimeter. Substitution of Eqs. (2.12) and (2.21) into (2.20) yields

$$(G^s\, d\mathscr{A})_{T,P} = \gamma\, d\mathscr{A} + \sum_i \mu_i c_i^s\, d\mathscr{A} \tag{2.22}$$

We may now divide by $d\mathscr{A}$ to obtain, after rearrangement,

$$\gamma = G^s - \sum_i \mu_i c_i^s \tag{2.23}$$

The differential of Eq. (2.23) gives

$$d\gamma = dG^s - \sum_i \mu_i\, dc_i^s - \sum_i c_i^s\, d\mu_i \tag{2.24}$$

Consider now the change in the surface free energy dG^s as a function of the experimental variables T, P, $\mathscr{A}$ and the surface concentration of the ith component in the surface c_i^s. The pressure P in Eq. (2.19) is the external pressure due to the force acting normal to the surface. For the flat surface we are considering here, it remains constant, $\partial G^s / \partial P = 0$ (the relationship between P and the radius of curvature of the surface will be derived in Section 2.7). In a manner similar to Eq. (2.19), the change in the excess surface free energy for a constant unit surface area [i.e., $(\partial G^s / \partial \mathscr{A}) = 0$] is given by

$$dG^s = -S^s \, dT + \sum_i \mu_i \, dc_i^s \tag{2.25}$$

Substitution of this equation into (2.24) for the surface free energy yields an expression for the change in surface tension as a function of temperature and surface composition:

$$d\gamma = -S^s \, dT - \sum_i c_i^s \, d\mu_i \tag{2.26}$$

This equation was first derived by Gibbs[11] and is commonly called the *Gibbs equation*. Just like the free-energy relations for bulk phases, it predicts the changes in surface properties that should occur as a function of the different experimental variables. Detailed discussion of the dependence of $d\gamma$ on the surface composition in a multicomponent system is outside the scope of this book. Excellent treatments of the surface thermodynamics of multicomponent systems of many types (liquid–gas, liquid–solid, liquid–liquid) are available in the literature.[2] One of the important pieces of information obtainable from the Gibbs equation is the variation of the surface tension γ with changes of the surface concentration c_2^s of the second component. To demonstrate its usefulness, we shall give here two examples of the use of the Gibbs equation for binary systems.

2.4.1. GAS ADSORPTION ON A LIQUID SURFACE

Consider the two-component system of liquid (component 1) and gas (component 2). The liquid is saturated with the gas at a temperature, T, which is held constant. If f_2 is the fugacity of the gas, then its chemical potential, μ_2, is given as

$$\mu_2 = \mu_2^0 + RT \ln f_2 \tag{2.27}$$

If we neglect the partial pressure of the liquid, the variation of the surface tension with the partial pressure of the gas is

$$dy = -c_2^s \, d\mu_2 = -c_2^s \frac{RT}{f_2} \, df_2 \qquad (2.28)$$

At low pressures f_2 equals P_2, where P_2 is the gas partial pressure. The surface concentration of the gas is thus given by

$$c_2^s = -\frac{P_2}{RT} \frac{dy}{dP_2} \qquad (2.29)$$

For example, the surface tension of water falls by about 0.1 dyne/cm when the air pressure is increased from 1 to 2 atm at 20°C. Assuming that $P_2 \approx 1.5$ atm, we have

$$c_2^s = \frac{1.5 \times 0.1}{8.3 \times 10^7 \times 293} \times 6 \times 10^{23} \approx 3.7 \times 10^{12} \text{ molecules/cm}^2$$

The surface concentration of the gas under these conditions changes by less than 1 per cent of a monolayer (one monolayer is $\sim 10^{15}$ molecules/cm²).

2.4.2. IDEAL BINARY SOLUTION

For this case the Gibbs equation may be expressed, at a constant temperature, as

$$dy = -c_1^s \, d\mu_1 - c_2^s \, d\mu_2 \qquad (2.30)$$

It can be shown[2] that the surface tension of component 1 in an ideal dilute solution is given by ·

$$y = y_1 + \frac{RT}{a} \ln \frac{X_1^s}{X_1^b} \qquad (2.31)$$

where y_1 is the surface tension of the pure component and a is the surface area occupied by 1 mole of component 1. We have assumed "perfect" behavior; i.e., the surface areas occupied by the molecules in the two different components are the same ($a_1 = a_2 = a$). X_1^s and X_1^b are the mole fractions of component 1 in the surface and in the bulk, respectively. We can write an expression of the form of Eq. (2.31) for component 2 as well. For the two-component system, γ may be eliminated by taking the ratio

$$\frac{X_1^s}{X_2^s} = \frac{X_1^b}{X_2^b} \exp\left[(\gamma_2 - \gamma_1)\frac{a}{RT}\right] \tag{2.32}$$

Equation (2.32) shows that the surface composition depends on the bulk composition and on the surface tension of the pure components. The surface will be richer in the component that has the smaller surface tension. In Fig. 2.2 the surface tension of isooctane-benzene is shown as a function of the isooctane mole fraction as determined by experiment. The agreement between the data and Eq. (2.32) is satisfactory.

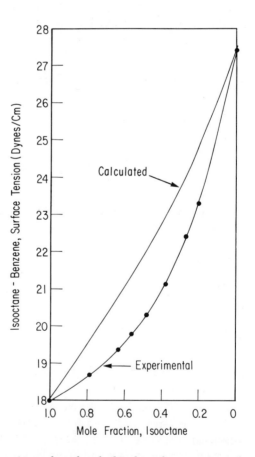

Figure 2.2. *Experimental and calculated surface tension of an isooctane–benzene solution as a function of changing mole fraction of isooctane.*

2.5 Theoretical Estimations of Specific Surface Energies

There have been several calculations carried out to estimate the specific surface energies and surface tensions for different crystal faces of ionic,[12] metal,[13] and noble gas crystals.[14] In these computations a suitable potential function is used that gives the potential energy of interaction between pairs of particles as a function of the distance of separation only. Then the total energy for the surface layer is obtained as the sum of the interactions of the pairs. For noble gas crystals the Lennard-Jones 6-12 potential function can be used for atoms i and j, which are separated by a distance r_{ij}:

$$\mathbf{V}_{ij} = -C_6 r_{ij}^{-6} + C_{12} r_{ij}^{-22} \tag{2.33}$$

Here C_6 and C_{12} are adjustable parameters. For ionic crystals the potential function used is

$$\mathbf{V}_{ij} = -e_i e_j r_{ij}^{-1} + b r_{ij}^{-n} \tag{2.34}$$

Here the first term gives the Coulombic interaction between ions of charges e_i and e_j, and b and n are adjustable parameters. In both Eqs. (2.33) and (2.34) the first terms represent attractive interaction, and the second terms give the repulsive parts of the potentials.

For calculating metal-surface properties a Morse potential function of the form

$$\mathbf{V}_{ij} = D(e^{-2\alpha r_{-ij}} - 2e^{-\alpha r_{-ij}}) \tag{2.35}$$

is used. Here, again, there are two adjustable parameters, D and α.

The computations in general are carried out in two parts. First a surface is "produced" by "breaking bonds," that is, by removing atoms adjacent to the newly created surface atoms. During this process the atoms in the newly created surface are held rigidly in equilibrium positions, which they would have occupied in the bulk of the solid. Then the surface atoms are allowed to "relax" into new equilibrium positions by displacement perpendicular to the surface plane. This relaxation process always decreases the total specific surface energy. Since atoms in the surface have fewer neighbors and the crystal symmetry is lowered with respect to atoms in the bulk, the displacement and subsequent lowering of the specific surface energy can be appreciable. The specific surface energy may be written as

$$E^s = E^s(0) + \Delta E^s \tag{2.36}$$

where $E^s(0)$ is the specific surface energy of the rigid lattice and ΔE^s is the relaxation energy. These two terms have opposite signs.

The results of these computations are very sensitive to the properties of the potential function used and to the atom or ion sizes assumed. For example, for ionic crystals the ions at the surface experience a net electric field that gives rise to polarization effects that will, in turn, have large effects on the relaxation energy. Because of these difficulties the computed values are only significant for indicating the order of magnitude of the specific surface energies and relaxation energies. Most of these calculations have not been verified by experiments. In Table 2.2 we list values of E^s, $E^s(0)$, ΔE^s, and γ for several noble gas, ionic, and metal crystal surfaces. Inspection of Table 2.2 indicates that the magnitudes of the specific surface energies for the different solids roughly parallel their heats of atomization; the weakly bonded rare gas crystals have low specific surface energies (20–60 ergs/cm²). Also, their specific relaxation energies are only a small fraction (<1 per cent) of the total surface energy. For singly charged ionic crystals the specific surface energies are almost an order of magnitude higher (100–300 ergs/cm²) than for rare-gas crystal surfaces. The relaxation energies are very high, 20–50 per cent of $E^s(0)$, because of the polarization effects mentioned above. For metals the specific surface energies are even higher (400–1000 ergs/cm²), but the relaxation energies are small—no more than 2–6 per cent of $E^s(0)$.

Another important result of these calculations is that to obtain minimum specific surface energies the surface layer had to be displaced outward, away from the second atomic plane by 2–8 per cent of the interplanar distance. Most of the computations predict lattice expansion at the surface. For ionic crystals the positive and negative ions in the surface may be displaced in opposite directions or in the same direction, depending on the relative ion sizes. These calculations also indicate that surface distortions are restricted to the proximity of the surface layer. The change in separation between atomic planes decreases as the inverse cube[14] or as the exponential[15] function of the distance from the surface.

We can also see from Table 2.2 that the specific surface energy varies appreciably from surface to surface. In general the surfaces with the highest atomic density have the lowest specific surface energy. These are the (111) face for face-centered cubic solids or the (110) face for body-centered cubic crystals. Also, the cleavage faces are expected to be surfaces of lowest surface energy [for example, the (100) face of alkali halide crystals].

Table 2.2. *Specific Surface Energies E^s, Specific Surface Energies of the Rigid Lattice $E^s(0)$, and Relaxation Energies ΔE^s Computed for Several Rare Gas, Alkali Halide, and Metal Surfaces*

Solid	E^s (ergs/cm²)	$E^s(0)$ (ergs/cm²)	ΔE^s (ergs/cm²)	Crystal face
Ne[24]	19.7	19.76	−0.06	(111)
	20.34	20.52	−0.18	(100)
Ar[24]	43.17	43.31	−0.14	(111)
	44.57	44.97	−0.40	(100)
Kr[24]	52.79	52.97	−0.18	(111)
	54.50	54.99	−0.49	(100)
Xe[24]	62.11	62.32	−0.21	(111)
	64.13	64.70	−0.57	(100)
LiF[23]	142	288.7	−146.4	(100)
	568	962.3	−394.6	(110)
LiCl[23]	107	251.4	−144.7	(100)
	340	599.2	−259.1	(110)
NaF[23]	216	265.9	−49.5	(100)
	555	711.7	−156.4	(110)
NaCl[23]	158	210.9	−52.7	(100)
	354	469.7	−115.6	(110)
KF[23]	184	225.9	−41.9	(100)
	423	528.0	−105.2	(110)
KCl[23]	141	175.3	−34.0	(100)
	298	367.3	−69.6	(110)
Cu[13]	3373.3	3440	−66.7	(111)
	3789	3980	−191	(100)
	5590	5820	−230	(110)
Al[13]	2931	3000	−69	(111)
	3221	3420	−193	(100)
	4762	5000	−238	(110)
Ag[13]	2537	2560	−23	(111)
	2895	2980	−85	(100)
	4360	4460	−100	(110)
Ni[13]	3971	4040	−69	(111)
	4442	4650	−208	(100)
	6586	6850	−264	(110)

2.6. Equilibrium Shape of a Crystal

In equilibrium the crystal will take up a shape that corresponds to a minimum value of the total surface free energy. Thus a stable crystal will be bounded primarily by crystal faces with the lowest surface tension since $G_{total}^s = G^s \mathscr{A} = \gamma \mathscr{A}$. In order to have the equilibrium shape, the integral $\int \gamma \, d\mathscr{A}$ over all surfaces of the crystal must be a minimum. If the specific surface free energies or surface tensions of the various crystal faces are known, one can construct the stable crystal shape using the method described by Wulff.[16] Let us define γ_i as the surface tension of the ith plane and $\mathbf{r}_i$ as a vector whose length is proportional to the surface tension of the ith plane (γ_i) and of direction normal to that crystal plane. From a common point one can draw a set of vectors whose lengths are proportional to the surface tension of each plane. Then one may construct a set of planes normal to each vector and positioned at its end; the envelope of these planes will define the equilibrium shape of the crystal.

Crystals rarely take up this equilibrium shape because they are, in general, prepared by nonequilibrium processes. However, if a small crystal is heated close to its melting point in equilibrium with its vapor, where atoms will have sufficient mobility to rearrange, the crystal may approach the equilibrium form. For such a crystal, using Wulff's construction, the relative magnitudes of the surface tensions of the different crystal faces may be deduced from the equilibrium shape.

Crystal faces that have high atomic density and are characterized by small Miller indices [(111), (100), etc.] have the lowest surface free energy and are, therefore, most stable. It would be important to determine the variation of the surface free energy as a function of crystal orientation between two low index crystal faces. Calculation[17] of the surface free energy, γ, as a function of angle in a given crystallographic zone [for example from the (111) face of a face-centered cubic crystal in the direction of the (110) face] shows that there are several surface free energy minima that correspond to stable surface structures. The plot of the surface free energy as a function of crystal orientation is called the γ-plot. It appears that surfaces with ordered steps of monatomic height correspond to surface structures with a surface free energy minimum (see for example Fig. 1.37).

The relative surface tension of different crystal faces may be determined experimentally by "boundary grooving" techniques.[17] The surface is chemically etched or scratched until a well-defined groove appears. Then, the groove slope is measured after annealing the surface by heating it to elevated temperature in a nonreactive atmosphere to allow the equilibration of surface atoms on the different crystal planes by surface diffusion. Assuming that the groove slope is due to the formation of well-defined terraces of surface tension γ, as shown in Fig. 2.3, the

Figure 2.3. *Schematic representation of the slope of a groove consisting of well-defined terraces with surface tension γ. The projection per unit area of slope is $\gamma/\cos\theta$.*

surface tension per unit area of slope is just $\gamma/\cos\theta$. Determination of the groove slope along different crystallographic orientations permits the construction of the γ plot, since it yields the ratio of surface tension along the different crystal faces. Typical surface-tension ratios are $[\gamma(111)/\gamma(100)]_{\text{Cu}} = 0.95$, $[\gamma(100)/\gamma(110)]_{\text{Cu}} = 0.91$, and $[\gamma(100)/\gamma(110)]_{\text{Mo}} = 1.14$.

2.7. The Internal and External Pressures at Curved Surfaces

Solids and liquids will always tend to minimize their surface area in order to decrease the excess surface free energy. For liquids, therefore, the equilibrium surface becomes curved where the radius of curvature will depend on the pressure difference on the two sides of the interface and on the surface tension. It should be remembered that we have considered the surface tension as a surface pressure which is exerted tangentially along the surface. Now we shall also consider the role of the external and internal pressures which act normal to the interface on the properties of the surface.

In Fig. 2.4 we have an equilibrium curved surface (a bubble in this case) with internal and external pressures P_{in} and P_{ext}, respectively, and with a surface tension γ. The radius of curvature of the bubble is given by r and hence its volume $V = \frac{4}{3}\pi r^3$. The force operating normal to the surface which would change the volume by $dV = 4\pi r^2 \, dr$ is given by $4\pi r^2(P_{in} - P_{ext}) \, dr$. The force exerted on the bubble surface tangentially is given by

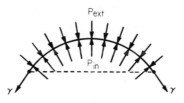

Figure 2.4. *Section of a bubble surface with internal and external pressures P_{in} and P_{ext} and surface tension γ.*

$4\pi r^2 \gamma$. If there is equilibrium, any change in the external pressure which would, for instance, increase the volume must be counteracted by a corresponding increase in the surface energy because of the extension of the surface by

$$\gamma \, d\mathscr{A} = 8\pi\gamma r \, dr \tag{2.37}$$

In equilibrium the two changes are equal and we have

$$4\pi r^2(P_{in} - P_{ext}) \, dr = 8\pi r\gamma \, dr \tag{2.38}$$

$$(P_{in} - P_{ext}) = \frac{2\gamma}{r} \tag{2.39}$$

This equation is quite significant in explaining the properties of liquid surfaces and of bubbles. First of all, Eq. (2.39) indicates that in equilibrium a pressure difference can be maintained across a curved surface. The pressure inside the liquid drop or gas bubble is higher than the external pressure, because of the surface tension. The smaller the droplet or larger the surface tension, the larger the pressure difference that can be maintained. For a flat surface $r = \infty$, and the pressure difference normal to the interface vanishes.

2.8. *The Vapor Pressure of Curved Surfaces*

Let us now consider how the vapor pressure of a droplet depends on its radius of curvature r. In equilibrium the pressure difference across the droplet is given by Eq. (2.39). If we transfer atoms from, say, the liquid to the surrounding gas phase, there is a small and equal equilibrium displacement on both sides of the interface:

$$dP_{in} - dP_{ext} = d\left(\frac{2\gamma}{r}\right) \tag{2.40}$$

At constant temperature the free-energy change associated with the transfer of atoms across the interface is given by

$$dG_{in} = V_{in}\, dP_{in} \tag{2.41}$$

and

$$dG_{ext} = V_{ext}\, dP_{ext} \tag{2.41a}$$

In equilibrium the free-energy changes are equal, dG_{in} equals dG_{ext}, and we have

$$V_{in}\, dP_{in} = V_{ext}\, dP_{ext} \tag{2.42}$$

Substitution of Eq. (2.40) into (2.42), after rearrangement, yields

$$\frac{V_{ext} - V_{in}}{V_{in}}\, dP_{ext} = d\left(\frac{2\gamma}{r}\right) \tag{2.43}$$

We can neglect the molar volume of the liquid with respect to the much larger molar volume of the gas ($V_{ext} \gg V_{in}$). Assuming that the vapor behaves as an ideal gas, we have for the molar volume

$$V_{ext} = \frac{RT}{P_{ext}} \tag{2.44}$$

Substitution of Eq. (2.44) into (2.43) yields, after rearrangement,

$$\frac{dP_{ext}}{P_{ext}} = \frac{2\gamma V_{in}}{RT}\, d\left(\frac{1}{r}\right) \tag{2.45}$$

We can now integrate Eq. (2.45) between the limits of a flat surface with zero curvature ($1/r = 0$, $P = P_0$) and some other state corresponding to a curved surface ($1/r$, P) and assume that the molar volume of the liquid remains unchanged along this path. We obtain

$$\ln\left(\frac{P}{P_0}\right) = \frac{2\gamma V_{in}}{RTr} \tag{2.46}$$

which is the well-known Kelvin equation for describing the dependence of the vapor pressure of any spherical particle on its size. We can see that, according to Eq. (2.46), small particles have higher vapor pressures than larger ones. Similarly, very small particles of solids have greater solubility than large particles. If we have a distribution of particles of different sizes, we will find that the larger particles will grow at the expense of the smaller ones, as predicted by Eq. (2.46).

It should be noted that such differences in vapor pressure or solubility that depend on particle size (radius of curvature) can only be observed for particles smaller than $r \leqslant 100$ Å. If we assume representative values for a water droplet ($\gamma = 73.4$ ergs/cm², $\bar{V}_{in} = 18$ cm³/mole), P/P_0 approaches unity rapidly above this radius.

2.9. The Contact Angle and Adhesion

The shape of the curved surface, in turn, allows one to determine the surface tension of the liquid when it is in equilibrium with its own vapor or to determine the "interfacial tension" if the droplet is in contact with a different substance (gas, liquid, or solid) instead of its own equilibrium vapor. The interfacial tension is determined by measuring the "contact angle" at the liquid–solid and solid–vapor interfaces. The contact angle is defined in Fig. 2.5, which shows a typical liquid–solid interface.

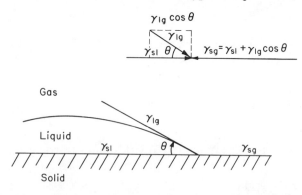

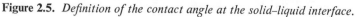

Figure 2.5. *Definition of the contact angle at the solid–liquid interface.*

Common experience tells us that the smaller the contact angle between the liquid and the solid, the more evenly is the liquid spread over or adheres to the solid surface, until at $\theta \approx 0°$ complete "wetting" of the solid surface takes place. If the contact angle is large ($\theta \rightarrow 90°$), the liquid does not readily wet the solid surface. For $\theta > 90°$ the liquid tends to form sphere-shaped droplets on the solid surface that may easily run off; i.e., the liquid does not wet the solid surface at all.

For a liquid that rests on a smooth surface with a finite contact angle, one can determine the relationship between the interfacial tensions at the different interfaces from consideration of the balance of surface forces at the line of contact of the three phases (solid, liquid, and gas). Remembering that the interfacial tension always exerts a pressure tangentially along the surface, the surface free-energy balance (a condition of equilibrium) between the surface forces acting in opposite directions is given by

$$\gamma_{lg} \cos\theta + \gamma_{sl} = \gamma_{sg} \tag{2.47}$$

or

$$\cos\theta = \frac{\gamma_{sg} - \gamma_{sl}}{\gamma_{lg}} \tag{2.47a}$$

Here γ_{lg} is the interfacial tension at the liquid–gas interface, γ_{sg} and γ_{sl} are the interfacial tensions between the solid–gas and solid–liquid interfaces, respectively. Thus in equilibrium at the solid–liquid–gas interface, from the knowledge of γ_{lg} and the contact angle, we can determine the *difference* $\gamma_{sg} - \gamma_{sl}$ but not their absolute values.

There are extreme cases when Eq. (2.47a) does not define the equilibrium position of the line of contact. This happens under conditions of complete wetting when $\gamma_{sg} > \gamma_{sl} + \gamma_{lg}$ or when the liquid does not wet the solid at all and the gas displaces the liquid completely along the solid surface $\gamma_{sl} > \gamma_{sg} + \gamma_{lg}$. Therefore, it is more convenient to consider the wetting coefficient

$$k_W = \frac{\gamma_{sg} - \gamma_{sl}}{\gamma_{lg}} \tag{2.48}$$

in describing the wetting ability of a liquid. If $k \geqslant +1$, the solid is completely wetted by the liquid. For k values between ± 1, the wetting is described by Eq. (2.47a), and for $k < -1$, the solid is not wetted at all. Since the wetting ability of the liquid at the solid surface is so important in practical problems of adhesion or lubrication, there is a

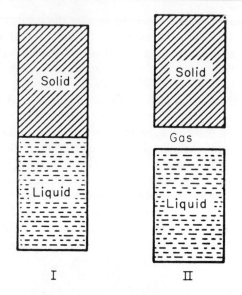

Figure 2.6. *Two states to define the reversible work of separation.*

great deal of work being carried out to determine the interfacial tensions for different combinations of interfaces.[5,18] It is, for example, useful in these studies to determine the energy necessary to separate the solid–liquid interface. In Fig. 2.6 the two states are shown before and after separation. We define the reversible work of separation as the difference in free energy between the two states in Fig. 2.6. In the process we eliminate the solid–liquid interface and re-form the solid–gas and liquid–gas interfaces. Thus the reversible work per unit area, W^s, is given by

$$W^s = \gamma_{lg} + \gamma_{sg} - \gamma_{sl} \tag{2.49}$$

This is called *Dupre's equation*. Combining Eqs. (2.47a) and (2.49) we have

$$W^s = \gamma_{lg}(1 + \cos\theta) \tag{2.50}$$

which is frequently called *Young's equation*. Finally, Harkins and his coworkers have defined the work of adhesion by redefining the second state (in Fig. 2.6) in which the solid and liquid phases are separated to be in a vacuum.[19] The surfaces in their separated state are free of adsorbed molecules; this is shown in Fig. 2.7. Thus the work of adhesion is defined as

$$W_A^s = \gamma_{l,0} + \gamma_{s,0} - \gamma_{sl} \tag{2.51}$$

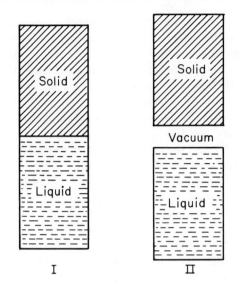

Figure 2.7. *Two states by Harkins to define the reversible work of separation.*

where $\gamma_{1,0}$ and $\gamma_{s,0}$ are the surface tensions of the liquid and the solid, respectively, in vacuum.

Surface roughness plays an important role in measurements of the contact angle.[5,18] Surface heterogeneities may modify the contact angle markedly with respect to that on a smooth solid surface.

Table 2.3 gives the range of values of the work of adhesion for several liquids on a variety of solids. In general, solids and liquids that have large surface tensions form strong adhesive bonds, i.e., have large negative works of adhesion. The work of adhesion ranges from 40 ergs/cm² for solid–liquid pairs that have low surface tension to as high as 140 ergs/cm² or greater for solid–liquid interfaces that have high

Table 2.3. *Work of Adhesion of Several Liquids on Most Solids*

Liquid	Work of adhesion[18] (ergs/cm²)
Water	60 to 140 on most solids
n-Octane	~ 46 on most solids
Benzene	50 to 60 on most solids
Nitrobenzene	70 to 90 on most solids
Glycerol	50 to 120 on most solids

surface tensions before joining them together.[18] Assuming that we have approximately 10^{13} molecules/cm^2 [since most adhesive liquids are polymers of large molecular areas ($\approx 5 \times 200$ Å)], the works of adhesion correspond to bond energies between the solid and the liquid on the order of 10^{-12}–10^{-11} erg/molecule. Application of the conversion factor, 1 cal/mole $= 7 \times 10^{-17}$ erg/molecule, gives works of adhesion on the order of $\geqslant 60$ kcal/mole. These energies per molecule are at least 100 times greater than the thermal energy per molecule, kT, at 300°K, indicating that adhesion should yield strong surface bonds. Adhesion of liquids to solids is always enhanced by high values of γ_{sg} (thus solid surfaces should be thoroughly cleaned) and low values of γ_{sl} (which indicates strong interaction between the solid and liquid surface atoms) according to Eq. (2.49).

There are several theories[20,21] and empirical correlation of experimental data that yield closed-form relationships, different from that in Eq. (2.47a), between the interfacial tensions γ_{sl}, γ_{lg}, and γ_{sg} and the contact angle θ. These are discussed in detail in several books on this subject.[4,5,18]

2.10. Nucleation

Since the free energy of formation of a surface is always positive, a particle that consists only of surfaces, i.e., platelets or droplets of atomic dimensions, would be thermodynamically unstable. This is also apparent from the Kelvin equation [Eq. (2.46)], which states that a particle that falls below a certain size will have an increased vapor pressure and therefore evaporate. There must be a stabilizing influence, however, that allows small particles of atomic dimensions to form and grow—a common occurrence in nature. This is given by the free energy of formation of the bulk condensed phase. In this process n moles of vapor are transferred to the liquid phase under isothermal conditions. This work of isothermal compression is given by

$$\Delta G = -nRT \ln \left(\frac{P}{P_{eq}} \right) \tag{2.52}$$

The condensed phase of n moles will grow if the vapor pressure P is larger than the equilibrium vapor pressure P_{eq}, or it will vaporize if $P < P_{eq}$. For small particles we have to take into account both their positive surface free energy, which would impede their growth, and the free energy of formation of the bulk condensed phase, which is negative for any external pressure larger than the equilibrium pressure. For a spherical droplet the total surface free energy is $4\pi r^2 \gamma$. Thus the total free energy of a growing particle is

$$\Delta G(\text{total}) = -nRT \ln \left(\frac{P}{P_{\text{eq}}}\right) + 4\pi r^2 \gamma \qquad (2.53)$$

In order to see the dependence of ΔG on the dimensions of the condensed particle, let us substitute for n, the number of moles of condensed vapor, $n = \frac{4}{3}\pi r^3 / V_m$, where $V_m = M/\rho$ is the molar volume of the condensed phase, M its atomic weight, and ρ the density. We have

$$\Delta G(\text{total}) = -\left(\frac{\frac{4}{3}\pi r^3}{V_m}\right) RT \ln \left(\frac{P}{P_{\text{eq}}}\right) + 4\pi r^2 \gamma \qquad (2.54)$$

Initially, when the condensed particle is very small, the surface free-energy term must be the larger of the two terms on the right-hand side of Eq. (2.54), and ΔG increases with r. In this range of sizes the particles are unstable. Above a critical size, however, the volumetric term becomes larger and dominates, since it increases as $\sim r^3$, but the surface free-energy term increases only as $\sim r^2$. Hence the particle of that size or larger grows spontaneously (for $P > P_{\text{eq}}$). When ΔG is at a maximum, i.e., $\partial \Delta G(\text{total})/\partial r = 0$, the particle reaches the critical size it must have for spontaneous growth to begin:

$$\frac{\partial \Delta G(\text{total})}{\partial r} = \frac{-4\pi}{V_m} r^2 RT \ln \left(\frac{P}{P_{\text{eq}}}\right) + 8\pi \gamma r = 0 \qquad (2.55)$$

Hence

$$r(\text{critical}) = \frac{2\gamma V_m}{RT \ln (P/P_{\text{eq}})} \qquad (2.56)$$

$r(\text{critical})$ is about 6–10 Å for most materials, which indicates that the droplet of critical size contains between 50 and 100 atoms or molecules. Figure 2.8 shows the variation of the total free energy with the radius of the condensed particle.

The free energy of a particle of critical size can be expressed by substituting Eq. (2.56) into (2.54). We obtain

$$\Delta G(\text{maximum}) = \frac{4\pi \gamma r_{\text{crit}}^2}{3} \qquad (2.57)$$

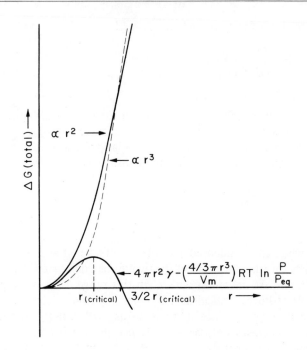

Figure 2.8. *Total free-energy change,* $\Delta G(total)$, *of a particle as a function of its radius r and the change of its surface and volume free energy as a function of r.*

Thus the total free energy for a spherical particle of critical size is one third of its surface free energy. Figure 2.9 shows the case of water at 0°C and its critical radius, which is about 8 Å.

A condensed particle must be larger than a certain critical size for spontaneous growth to occur at pressures $P > P_{eq}$. "Homogeneous nucleation" of the condensed phase by simultaneous clustering of many vapor atoms to reach this critical size, however, is very improbable. This is the reason "supersaturated vapor" can exist; i.e., ambient conditions in which vapor pressures are larger than the equilibrium vapor pressure ($P > P_{eq}$) of a condensible substance can be established without the formation of the condensed phase. Precipitation in the absence of nuclei is very difficult, and large pressures much higher than the equilibrium vapor pressure ($P \gg P_{eq}$) must be established before condensation could occur within reasonable experimental times. There is, in general, a long "induction period" before growth of a liquid or solid phase commences in the absence of stable nuclei. Supersaturated vapor or undercooled liquid (cooled many degrees below its freezing point) are common occurrences in nature and in the laboratory.

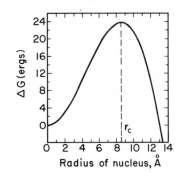

Figure 2.9. *Total free-energy change of a water droplet as a function of its radius r.*

Because of the difficulty of homogeneous nucleation, growth of condensed phases generally occur on solid surfaces already present, such as the walls of the reaction cell or dust particles present in the atmosphere or in interstellar space. The introduction of solid crystallites to induce the formation and growth of the condensed phase is frequently called "seeding" or "heterogeneous nucleation." It is used to facilitate the growth of crystals or the condensation of water droplets from the atmosphere (rain making).

Summary

The surface thermodynamic functions are defined as excesses of the bulk thermodynamic properties that are due to the presence of the surface surrounding the condensed phase. Creation of a stable surface always has a positive free energy of formation; therefore solids or liquids tend to minimize their surface area by forming crystal planes of lowest free energy or by assuming a spherical shape, respectively. The change in surface tension as a function of surface composition and temperature is expressed by the Gibbs equation. Small particles ($r \leqslant 100$ Å) have higher vapor pressure or solubility than larger ones due to the difference between the internal and external pressures at curved surfaces. The contact angle between the liquid and the solid reveals the relationship between the surface forces at the line of contact that determine adhesion, lubrication or wetting. A balance between the positive surface free energy and the negative free energy of formation of the condensed phase gives rise to the nucleation phenomenon that has many consequences in nature (supersaturation, undercooling and the predominance of heterogeneous nucleation).

References

1. G. N. Lewis and M. Randall: *Thermodynamics*, revised by K. S. Pitzer and L. Brewer. McGraw-Hill Book Company, New York, 1961.
2. R. Defay and I. Prigogine: *Surface Tension and Adsorption*. John Wiley & Sons, Inc., New York, 1966.
3. J. C. Erickson: *Surface Sci.*, **14**, 221 (1969).
4. A. W. Adamson: *Physical Chemistry of Surfaces*. John Wiley & Sons, Inc. (Interscience Divison), New York, 1967.
5. J. J. Bikerman: *Surface Chemistry for Industrial Research*. Academic Press, Inc., New York, 1947.
6a. L. L. Bircumshaw: *Phil. Mag.*, **17**, 181 (1934).
6b. A. F. Crawley, H. R. Thresh, and D. W. G. White: Report PN-67-18, Department of Engineering, Mines and Resources, Ottawa, Canada.
7. E. A. Guggenheim: *J. Chem. Phys.*, **13**, 253 (1945).
8. A. Ferguson: *Trans. Faraday Soc.*, **19**, 408 (1923).
9. D. W. G. White: *Trans. Met. Soc. AIME*, **236**, 796 (1966).
10. P. Balk and G. C. Benson: *J. Phys. Chem.*, **63**, 1009 (1959).
11. J. W. Gibbs: *The Collected Work of J. W. Gibbs*. Longmans, Green & Co. Ltd., London, 1931.
12. G. C. Benson: *J. Chem. Phys.*, **35**, 2113 (1961); **39**, 302 (1963).
13. J. J. Burton and G. Jura: *J. Phys. Chem.*, **71**, 1937 (1967).
14. B. J. Alder, J. R. Vaisnys, and G. Jura: *J. Phys. Chem. Solids*, **11**, 182 (1959).
15. B. C. Clark, R. Herman, and R. F. Wallis: *Phys. Rev.*, **139**, A860 (1965).
16. G. Wulff, *Z. Krist.*, **34**, 449 (1901).
17a. W. L. Winterbottom: in *Surfaces and Interfaces*, J. J. Burke et al., eds. Syracuse University Press, Syracuse, N.Y., 1967.
17b. C. Herring: in *Structure and Properties of Solid Surfaces*, R. Gomer and C. S. Smith, eds. University of Chicago Press, Chicago, Ill., 1962.
17c. N. Cabrera and R. V. Coleman: in *The Art and Science of Growing Crystals*, J. J. Gilman, ed. John Wiley and Sons, Inc., New York, 1963.
18. D. J. Alner, ed.: *Aspects of Adhesion*. University of London Press Ltd., London, 1966.
19. W. D. Harkins: *The Physical Chemistry of Surface Films*. Van Nostrand Reinhold Company, New York, 1952.
20. H. W. Fox and W. A. Zisman: *J. Colloid Sci.*, **7**, 428 (1952).
21. R. J. Good and L. A. Girifalco: *J. Phys. Chem.*, **61**, 944 (1957); **64**, 561 (1960).
22. J. M. Blakely and P. S. Maiya: in *Surfaces and Interfaces*, J. J. Burke et al., eds. Syracuse University Press, Syracuse, N.Y., 1967.
23. G. C. Benson and R. S. Yuen: in *The Solid-Gas Interface*, E. A. Flood, ed. Marcel Dekker, Inc., New York, 1967.
24. G. C. Benson and T. A. Claeton: *J. Phys. Chem. Solids*, **25**, 367 (1964).

Problems

2.1. Calculate the reversible work necessary to create an additional 1-mm² surface area of Ag(liquid), Ag(solid), NaCl(solid), and N_2(liquid).

2.2. The specific surface entropy of water is 0.155 erg/deg cm² [*J. Phys. Chem.*, **74**, 3024 (1970)]. The surface tension and the specific surface entropy of water are reduced upon the adsorption of *n*-butane (h-C_4H_{10}) and perfluoro-tributylamine [$(C_4F_9)_3N$]. Give reasons for the large reductions of $\gamma(H_2O)$ and $S^s(H_2O)$. Suggest five liquids or gases which would result in the reduction of $S^s(H_2O)$ for the same chemical reasons.

2.3. A researcher decides to measure the surface tension of Cu(liquid). However, some of his samples are contaminated by silver and some of them by nickel. How would these impurities affect his measurements? Could he, by comparing his results with those obtained for pure copper samples, determine the impurity concentration, and how? If he finds that the impure liquid shows a decreasing specific surface entropy in a given temperature range while the pure molten metal does not, what conclusion could he reach about the behavior of the two different types of contaminants as a function of temperature?

2.4. For an external pressure of 1 atm, calculate the internal pressure of a water droplet of $r = 50$ Å, 5×10^4 Å, and 5×10^6 Å.

2.5. Compute the vapor pressure of a spherical platinum particle of $r = 10$ Å at 200°C and at 800°C using the vapor pressure data given in the literature [for example, *J. Phys. Chem.*, **64**, 1323 (1960)]. Small particles of comparable size are frequently used as catalysts in many industrial processes. Could these particles be used continuously for three years assuming that the loss of material would only occur by vaporization?

2.6. The contact angle of liquid aluminum on α-alumina (Al_2O_3) surfaces is $\theta = 90°$ at 1000°C. The contact angle decreases with increasing temperature and it is $\Theta = 60°$ at 1200°C. If $\gamma_g(Al) = 750$ ergs/cm² at 1100°C, compute $\gamma_{sg} - \gamma_{sl}$ at 1100°C assuming that cos Θ varies linearly with temperature. [Ref.: *J. Am. Ceram. Soc.*, **51**, 569 (1968).]

2.7. Compute the ΔG(total) versus r curve for benzene from $r = 5$ Å to $r = 500$ Å for $(P/P_{eq}) = 10$ and 10^3.

2.8. It has been found that clean surfaces of like metals "stick" together when brought into contact in high vacuum. However, "dirty" metal surfaces that are covered by a layer of oxide do not adhere. Give an explanation of this effect.

2.9. The lubrication of sliding solid surfaces is one of the important problems in many technologies. The removal of unwanted deposits from solid and liquid surfaces is the major concern of many chemical industries. Suggest several ways to utilize the concepts of surface thermodynamics that were discussed in this chapter to develop better lubricants and detergents. Select practical systems (liquid or solid lubricants of metal contacts, oil spill in seawater, etc.), review the present state of the art, and recommend alternative methods of lubrication or chemical cleaning.

3

Dynamics of

Surface Atoms

3.1. Harmonic Oscillator Models of
Atomic Vibrations in Solids

IN STUDYING THE MOTION of surface atoms or molecules it is
frequently very useful to relate them to the motion of model systems that
have been utilized in other fields. Perhaps one of the most useful analogues
which is often used in chemistry and physics, is the harmonic oscillator—a
mass on a spring that is tied on one side (one-dimensional harmonic
oscillator) or tied down on all sides (three-dimensional harmonic
oscillator.) The properties of the simple one-dimensional harmonic oscillator can be directly related, in the first approximation, to the
properties of vibrating atoms in a molecule or in a solid. The
reason for the success of this simple model in exploring the dynamics of

atomic motion in solids and in surfaces is that the displacement of the vibrating atoms is only a small fraction of their interatomic distance. For small displacements the atomic vibrations can be well approximated to the vibrations of a harmonic oscillator. We will see that, using the harmonic oscillator model, we can get close to a description of the motion of surface atoms about their equilibrium position.

The restoring force f which the spring exerts on the displaced mass m is linearly proportional to the amount of stretch or displacement x and opposite in sign to it:

$$f = m\left(\frac{d^2x}{dt^2}\right) = -Kx \tag{3.1}$$

where d^2x/dt^2 is the acceleration and K is the constant of proportionality called the "force constant." This is illustrated in Fig. 3.1: r is the relaxed length of the spring at which $x = 0$, and $x = a$ is the distance of maximum displacement of the mass from its equilibrium position at r. Equation (3.1) has a solution of the form

$$x = a \cos \omega t \tag{3.2}$$

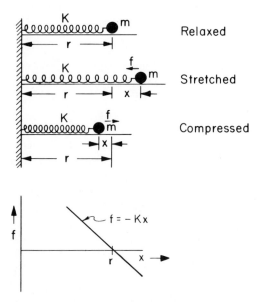

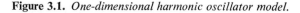

Figure 3.1. *One-dimensional harmonic oscillator model.*

where $\omega = (K/m)^{1/2}$ is the angular frequency, which depends on the mass m and the "stiffness" of the spring K, which is the force constant; a is the amplitude of oscillation, which is equal to the maximum displacement; and t is the time.

The potential-energy change $d\mathbf{V}$ of the harmonic oscillator while displacing the particle by dx against the force the spring exerts on the particle is

$$d\mathbf{V} = -f\,dx$$

or

$$\frac{d\mathbf{V}}{dx} = Kx \qquad (3.3)$$

Using the boundary condition that in its equilibrium position the mass makes zero displacement ($x = 0$) and has zero potential energy ($V = 0$), we have, by integrating Eq. (3.3),

$$\mathbf{V} = \tfrac{1}{2}Kx^2 \qquad (3.4)$$

The kinetic energy of the particle, $\mathbf{T}$, is given by $\tfrac{1}{2}m(dx/dt)^2$ or $p^2/2m$, where dx/dt is the velocity and p is the momentum. The total energy of the system, E, is then equal to the sum of $\mathbf{V}$ and $\mathbf{T}$:

$$E = \mathbf{V} + \mathbf{T} \qquad (3.5)$$

Substitution of Eq. (3.2) into the potential-energy and kinetic-energy expressions gives for the total energy,

$$E = \tfrac{1}{2}Ka^2 \cos^2 \omega t + \tfrac{1}{2}Ka^2 \sin^2 \omega t = \tfrac{1}{2}Ka^2 \qquad (3.6)$$

The potential energy is at a maximum where the displacement is the largest and diminishes as the atom approaches its equilibrium position ($x = 0$). On the other hand, the behavior of the kinetic-energy term is just the opposite. T is zero for maximum displacement [$(dx/dt)_{x=a} = 0$] and reaches its maximum value at $x = 0$. It should be noted that the total energy depends on the square of the amplitude. Thus a small change in the amplitude can greatly change the total energy. This consideration becomes important when we compare the amplitudes of vibration of surface atoms with those of the bulk atoms. It is also important to recognize that differentiation of Eq. (3.4) twice yields $d^2\mathbf{V}/dx^2 = K$. Thus the force constant of the spring is equal to the curvature of the potential-energy function, which is shown in Fig. 3.2.

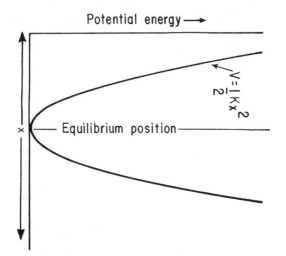

Figure 3.2. *Potential energy V of the one-dimensional harmonic oscillator as a function of the displacement x.*

In the case of an arbitrary (not necessarily harmonic) attractive potential, the potential-energy term can be expanded in a Taylor series about its minimum $(x = 0)$ to give

$$\mathbf{V}(x) = \mathbf{V}_{x=0} + \left(\frac{d\mathbf{V}}{dx}\right)_{x=0} x + \frac{1}{2}\left(\frac{d^2\mathbf{V}}{dx^2}\right)_{x=0} x^2 + \frac{1}{6}\left(\frac{d^3\mathbf{V}}{dx^3}\right)_{x=0} x^3 + \cdots \quad (3.7)$$

Since $\mathbf{V}_{x=0} = 0$ and the first derivative must also be zero because the potential energy is at a minimum at $x = 0$, the potential energy is given by the second- and higher-order derivatives in the series. In the "harmonic" approximation one neglects the higher-order terms, and this is adequate for our purposes at the present. By taking the third derivative into consideration we can describe the "anharmonicity" of the oscillator. The anharmonic nature of the real crystal potential is responsible for the thermal expansion of solids at elevated temperatures and for the interaction between certain modes of lattice vibration. For surface atoms placed in an anisotropic environment the inclusion of the anharmonic term in the potential-energy function could be especially important in describing many of the physical–chemical properties of the atoms.

We would now like to compute the average energy of a set of harmonic oscillators. We will do this by assuming that their energy distribution is continuous and obeys Boltzmann statistics. It is also assumed that the total energy of N such oscillators is the sum of the energy of the individual oscillators vibrating independently. Thus we are considering "classical" harmonic oscillators. The probability $\mathbf{P}$ that an oscillator will be found in a particular state i which has energy E_i is given by

$$\mathbf{P} = \frac{g_i e^{-E_i/k_B T}}{\sum\limits_{i=0} g_i e^{-E_i/k_B T}} \tag{3.8}$$

where k_B is the Boltzmann constant. The quantity g_i is the multiplicity or statistical weight of the energy state i; in this case $g_i = 1$. It gives the number of energy states that have the same energy. The sum in the denominator is the sum over all energy states and is called the *partition function*. The average energy of each oscillator can be found by multiplying the probability by the total energy of a given oscillator $[(p_i^2/2m) + m\omega^2 x_i^2/2)]$ and then summing over all energy states[1]:

$$\bar{E} = \frac{\sum\limits_{i=0}^{\infty} E_i e^{-E_i/k_B T}}{\sum\limits_{i=0}^{\infty} e^{-E_i/k_B T}} \tag{3.9}$$

The partition function $Z = \sum_{i=0}^{\infty} e^{-E_i/k_B t}$ in the denominator of Eq. (3.9) is related to the numerator of Eq. (3.9) in a simple manner. If we write $b = 1/k_B T$, the derivative of Z with respect to b is

$$\frac{\partial Z}{\partial b} = -\sum_{i=0}^{\infty} E_i e^{-E_i b}$$

which is the negative of the numerator. In general,

$$\bar{E} = \frac{-\dfrac{\partial}{\partial b} Z}{Z} = -\frac{\partial}{\partial b} \ln Z \tag{3.10}$$

This relationship is valid no matter what the expression for E_i may be, provided that it is independent of $k_B T$. For one-dimensional harmonic oscillators the partition function factors as follows:

$$Z = \sum_{i=0}^{\infty} \exp\left(\frac{-p_i^2}{2mk_B T}\right) \sum_{i=0}^{\infty} \exp\left(\frac{-x_i^2 m\omega^2}{2k_B T}\right) \tag{3.11a}$$

Since the energy distribution is continuous the summation can be replaced by integrals,

$$Z = Z_{kinetic} \times Z_{potential}$$

$$= \int_{-\infty}^{\infty} \exp -\left(\frac{b}{2m}\right) p^2 \, dp \int_{-\infty}^{\infty} \exp \left[-\left(\frac{m\omega^2 b}{2}\right) x^2\right] dx \qquad (3.11b)$$

Remembering the definite integral $\int_{-\infty}^{\infty} e^{-\alpha^2 x^2} \, dx = \sqrt{\pi}/\alpha$,

$$Z = \sqrt{2\pi m} \, b^{-1/2} \times \sqrt{\frac{2\pi}{m\omega^2}} \, b^{-1/2}$$

Then

$$\frac{\partial Z}{\partial b} = -\frac{1}{2}\sqrt{2\pi m} \, b^{-3/2} Z_{potential} - \frac{1}{2}\sqrt{\frac{2\pi}{m\omega^2}} \, b^{-3/2} Z_{kinetic}$$

$$= -\frac{1}{2b} Z_{kinetic} Z_{potential} - \frac{1}{2b} Z_{kinetic} Z_{potential} \qquad (3.12)$$

Finally,

$$\bar{E} = \frac{-\partial Z/\partial b}{Z} = \tfrac{1}{2}k_B T + \tfrac{1}{2}k_B T = k_B T \qquad (3.13)$$

For N number of three-dimensional harmonic oscillators we have $3N$. modes of vibration.[2] Thus the total energy E is

$$E = 3N\bar{E} = 3Nk_B T \qquad (3.14)$$

The heat capacity at constant volume, C_V, is given by the temperature derivative of the total energy (as was discussed in Chapter 2). Thus from Eq. (3.14) we have

$$C_V = \left(\frac{\partial E}{\partial T}\right)_V = 3Nk_B \qquad (3.15)$$

Therefore, the heat capacity of a set of independent harmonic oscillators with a continuous energy distribution is independent of temperature. This harmonic oscillator model gave adequate description of heat capacity of monatomic solids at elevated temperature. However, heat-capacity measurements that were extended to low temperatures during the last decades of the nineteenth century have indicated that the heat capacity of monatomic solids decreases with decreasing temperature. This discrepancy between the theory, which describes the motion of atoms in a solid as harmonic oscillators, and experiment was partially eliminated by Einstein's theory[3] of specific heats, which was proposed in 1906. He introduced the idea of quantized lattice vibrations.

Let us restrict the allowed frequencies of oscillation of the harmonic oscillator by quantizing the energy that can be stored in each vibration. That is, in order to oscillate, the "particle" must have at least one quantum of energy, $h\nu$, where h is Planck's constant and ν is the oscillation frequency, $\nu = \omega/2\pi$. Thus oscillation occurs only with frequencies that are integral multiples of ν. The total energy stored in a harmonic oscillator is given by[†]

$$E = nh\nu \tag{3.16}$$

where n is an integer. The average energy of the quantized harmonic oscillator is given by

$$\bar{E} = \frac{\sum_{n=0}^{\infty} nh\nu \exp\left(-nh\nu/k_B T\right)}{\sum_{n=0}^{\infty} \exp\left(-nh\nu/k_B T\right)} \tag{3.17}$$

in the same way as the average energy is given by Eq. (3.9) for harmonic oscillators with continuous energy distribution. However, one cannot integrate instead of summing over all states, since the energy distribution is quantized (discontinuous). In order to simplify Eq. (3.17) we make the substitution $x = -h\nu/k_B T$ and write out the terms in both the numerator and the denominator:

$$\bar{E} = h\nu \frac{e^x + 2e^{2x} + 3e^{3x} + \cdots}{1 + e^x + e^{2x} + e^{3x} + \cdots} \tag{3.18}$$

We can see that in this equation the numerator is again the derivative

---

[†] From quantum mechanics the energy of a harmonic oscillator is given by $E = (n + \frac{1}{2})h\nu$, where $E_0 = \frac{1}{2}h\nu$ is the zero-point energy. This implies that even at absolute zero the atoms in the crystal are still vibrating and store a rather large amount of energy. However, for calculations of the specific heat, which is the temperature derivative of the energy, and its temperature dependence, this constant term, which was not included in Einstein's original calculations, can be neglected since it vanishes under differentiation with respect to temperature.

of the denominator. Letting u be equal to the denominator, we have

$$\bar{E} = h\nu \frac{1}{u} \frac{du}{dx} = h\nu \frac{d}{dx} \ln u \tag{3.19}$$

Thus we can rewrite Eq. (3.18) as

$$\bar{E} = h\nu \frac{d}{dx} \ln (1 + e^x + e^{2x} + e^{3x} + \cdots) \tag{3.20}$$

Fortunately, the infinite sum can be written in closed form since

$$\frac{1}{1 - y} = 1 + y + y^2 + y^3 + \cdots \tag{3.21}$$

Substitution of Eq. (3.21) into (3.20) and subsequent differentiation yields the average energy:

$$\bar{E} = h\nu \frac{e^x}{1 - e^x} = \frac{h\nu}{e^{h\nu/k_B T} - 1} \tag{3.22}$$

Excitations of vibrations to different frequencies obey the Bose–Einstein statistics; i.e., on an average the number of quanta per oscillator[4] is

$$\bar{n} = \frac{\bar{E}}{h\nu} = \frac{1}{e^{h\nu/k_B T} - 1} \tag{3.23}$$

at any given temperature T. The quanta of oscillation are frequently called "phonons" if the term is used in discussing the vibrations of atoms in a solid. The total energy of N harmonic oscillators is then

$$E = 3N\bar{E} = \frac{3Nh\nu}{e^{h\nu/k_B T} - 1} \tag{3.24}$$

At high temperatures, $h\nu \ll k_B T$, we have

$$\exp\left(\frac{h\nu}{k_B T}\right) \approx 1 + \frac{h\nu}{k_B T} \tag{3.25}$$

so

$$E_{T \to \infty} \approx 3Nk_B T \tag{3.26}$$

This is the same result we have obtained by assuming continuous energy distribution. At low temperatures, however, $\exp(h\nu/k_BT) \gg 1$; therefore,

$$E_{T \to 0} \approx 3Nh\nu \exp\left(\frac{-h\nu}{k_BT}\right) \tag{3.27}$$

Thus the total energy of the system and its heat capacity as well should fall off exponentially at very low temperatures.

Let us now pick a single frequency of oscillation, ν_E, which is assumed to be characteristic of a given material and is defined by

$$h\nu_E = k_B\Theta_E \tag{3.28}$$

where ν_E and Θ_E are called the *Einstein frequency* and *Einstein temperature*, respectively. Equation (3.24) can then be rewritten as

$$E = \frac{3Nk_B\Theta_E}{e^{\Theta_E/T} - 1} \tag{3.29}$$

The heat capacity is given by

$$C_V = \left(\frac{\partial E}{\partial T}\right)_V = 3Nk_B\left(\frac{\Theta_E}{T}\right)^2 \frac{e^{\Theta_E/T}}{(e^{\Theta/T} - 1)^2} \tag{3.30}$$

By using Eq. (3.30) the Einstein model can be compared with the experimental data. With careful selection of Θ_E, the model fits most of the data down to a temperature as low as $0.2\Theta_E$. However, below this temperature (roughly below $10°K$) the experimental specific heats decrease consistently slower than predicted by the experimental temperature dependence expressed in the model. First it was thought that the model could be improved by assuming that the solid atoms can vibrate at two characteristic frequencies. Accurate specific-heat measurements have shown, however, that at these low temperatures the specific heat is proportional to the third power of the temperature: $C_V \propto T^3$. Finally, Debye[5] has proposed a model that explains this low-temperature behavior satisfactorily.

He considered a system of harmonic oscillators, which instead of vibrating at a characteristic frequency ν_E has a distribution of frequencies $g(\nu)$. This can come about if the harmonic oscillators interact with each other; i.e., the vibration becomes coupled because of interaction among neighbors. Debye has assumed that the solid behaves as a three-dimensional elastic continuum.[6,7] It can be shown [see, for example,

T. L. Hill, *Introduction to Statistical Thermodynamics* (Addison-Wesley Publishing Company, Inc., Reading, Mass., 1962), pp. 492–494] that for a three-dimensional isotropic solid continuum of volume V the number of frequencies, n, in the frequency range between ν and $\nu + d\nu$ is given by

$$\frac{dn}{d\nu} = g(\nu) = \frac{4\pi V \nu^2}{C_3} \tag{3.31}$$

which can be rewritten simply as

$$g(\nu) = B\nu^2 \tag{3.32}$$

where B is a constant related to the velocity of sound propagation in the solid, C_3,† and to the volume V. In Fig. 3.3 we can see the different

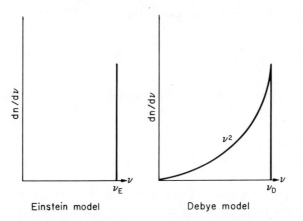

Figure 3.3. *Frequency distributions of lattice vibrations assumed by the Einstein and Debye models.*

frequency distributions that were assumed in the Einstein and Debye models. Another assumption by Debye was that there is a cutoff frequency ν_D that is the upper limit of the vibration frequencies of the coupled harmonic oscillators. Using these conditions, we have

$$\int_0^{\nu_D} B\nu^2 \, d\nu = 3N = \frac{B\nu_D^3}{3} \tag{3.33}$$

† The sound velocity in the elastic continuum has actually one longitudinal, C_l, and two transverse components, $2C_t$. Thus C_3 should actually be given by $C_3^{-1} = C_l^{-3} + 2C_t^{-3}$. In two dimensions, which we will consider later, we have $C_2^{-1} = C_l^{-2} + 2C_t^{-2}$.

since there are $3N$ modes of vibration for N atoms. From Eq. (3.33), $B = 9N/v_D^3$. Therefore, Eq. (3.32) can be rewritten as

$$g(v) = 9N \frac{v^2}{v_D^3} \tag{3.34}$$

Since the frequency distribution is continuous, we can integrate over all frequencies up to the cutoff "Debye frequency" v_D. The total energy of a system of coupled three-dimensional harmonic oscillators can then be expressed as

$$E = \int_0^{v_D} \frac{hvg(v)}{e^{hv/k_B T} - 1} \, dv \tag{3.35}$$

Then, by substitution of Eq. 3.34 to Eq. 3.35,

$$E = \frac{9Nh}{v_D^3} \int_0^{v_D} \frac{v^3}{e^{hv/k_B T} - 1} \, dv \tag{3.36}$$

Using the approximation to the exponential at high temperatures ($kT \gg hv$) given by Eq. (3.25), we have

$$E_{T \to \infty} = \frac{9Nk_B T}{v_D^3} \frac{v_D^3}{3} = 3Nk_B T \tag{3.37}$$

Thus the Debye frequency distribution gives the same high-temperature classical limit as predicted by the Einstein model of independent harmonic oscillators. At low temperatures ($hv \gg k_B T$) we have to use a simple mathematical trick to evaluate the integral easily. Let us make the substitution to a dimensionless function, $x = hv/k_B T$, in Eq. (3.36):

$$E = \frac{9Nk_B^4 T^4}{h^3 v_D^3} \int_0^{x_{\max}} \frac{x^3 \, dx}{e^x - 1} \tag{3.38}$$

where $x_{\max}$ is defined by $x_{\max} = hv_D/k_B T$. Θ_D is the characteristic Debye temperature that corresponds to the Debye cutoff frequency v_D by definition, $hv_D = k_B \Theta_D$. Now instead of cutting off the integral at $x = x_{\max}$, we extend the upper limit to $x = \infty$. The change of boundary conditions does not introduce any errors since at low temperatures none of the higher-frequency vibrational modes are excited.

In Fig. 3.4 there is a plot of the function $x^3/(e^x - 1)$ as a function of x which clearly shows that the error introduced at low temperatures by

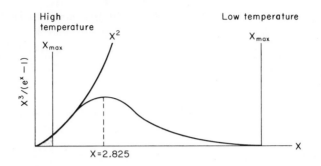

Figure 3.4. *Plot of the function $x^3/(e^x - 1)$ as a function of x.*

changing the limit is negligible. If the limits are from zero to infinity, however, we have the Riemann zeta function[8]

$$\int_0^\infty \frac{x^{s-1}\, dx}{e^x - 1} = (s-1)! \sum_{n=1}^\infty \frac{1}{n^s} \tag{3.39}$$

which has the value $\pi^4/15$ for $s = 4$. Thus

$$E_{T\to 0} = \frac{9Nk_B^4 T^4}{h^3 \nu_D^3} \int_0^\infty \frac{x^3\, dx}{e^x - 1} = \frac{9Nk_B^4 T^4}{h^3 \nu_D^3} \frac{\pi^4}{15} \tag{3.40}$$

Substituting for $\nu_D = k\Theta_D/h$ in Eq. (3.40), we have

$$E_{T\to 0} = \frac{9Nk_B \pi^4}{15\Theta_D^3} T^4 \tag{3.41}$$

Thus instead of the exponential temperature dependence of the total energy at low temperatures, which is predicted by the Einstein model, the Debye model predicts a T^4 power dependence. It is not the total energy of the solid but its temperature derivative (the heat capacity) that is measured by experiments. By differentiating the high- and low-temperature limits of the total energy, Eqs. (3.37) and (3.41), we have

$$C_V(T \to \infty) = 3Nk_B \tag{3.42}$$

and

$$C_V(T \to 0) = \tfrac{12}{5} Nk_B \pi^4 \left(\frac{T}{\Theta_D}\right)^3 = 234 Nk_B \left(\frac{T}{\Theta_D}\right)^3 \tag{3.43}$$

Thus the Debye model of lattice atom vibration correctly predicts the T^3 power dependence for the vibrational heat capacity of solids at low temperatures $(T < \Theta_D/10)$, which was found by accurate calorimetric studies.

3.2. Surface Heat Capacity

The use of the Debye model, which assumes that a solid behaves as a three-dimensional elastic continuum with a frequency distribution $g(v) = Bv^2$, allows accurate prediction of the temperature dependence of the vibrational heat capacity of solids at low temperatures $(C_V \propto T^3)$ as well as at high temperatures $(C_V = 3Nk_B)$. One may also use the same model with confidence to evaluate the temperature dependence of the surface heat capacity due to vibrations of atoms in the surface.

Let us now consider a surface of area $\mathscr{A}$ at the termination of the bulk lattice in which the atoms have the same properties as in the three-dimensional elastic continuum we have considered before. The number of frequencies, n, in the range v and $v + dv$ is given by[7]

$$\frac{dn}{dv} = g(v) = \frac{2\pi \mathscr{A} v}{C_2} \tag{3.44}$$

where C_2 is the sound velocity in two dimensions (see page 91).

In general, for an elastic continuum of μ dimensions, the frequency distribution is $g(v) \approx v^{\mu-1}$. If the N number of surface atoms still have $3N$ vibrational modes (they are allowed to have out-of-plane vibrations), we have

$$\int_0^{v_D} \frac{2\pi \mathscr{A}}{C_2} v \, dv = 3N = \frac{\pi \mathscr{A} v_D^2}{C_2} \tag{3.45}$$

Therefore, substitution of Eq. (3.45) into (3.44) yields

$$g(v) = \frac{6Nv}{v_D^2} \tag{3.46}$$

the total energy of the surface is given by

$$E = \frac{6Nh}{v_D^2} \int_0^{v_D} \frac{v^2}{e^{hv/k_BT} - 1} \, dv \tag{3.47}$$

Using the approximation for high temperatures, $k_B T \gg h\nu$ [$e^{h\nu/k_B T} \approx 1 + (h\nu/k_B T)$], we obtain

$$E_{T \to \infty} = \frac{6Nk_B T}{\nu_D^2} \frac{\nu_D^2}{2} = 3Nk_B T \qquad (3.48)$$

and

$$C_V(T \to \infty) = 3Nk_B$$

Thus the total energy and the heat capacity for surfaces yield the same limiting value as that for three-dimensional solids. At low temperatures, following the same procedure we have used in computing the total energy and heat capacity in three dimensions, we have, after substitution ($x = h\nu/k_B T$, $x_{\max} = h\nu_D/k_B T$, and $\nu_D = k_B \Theta_D/h$) into Eq. (3.47),

$$E = \frac{6Nk_B}{x_m^2 \Theta_D^2} T^3 \int_0^{x_m} \frac{x^2\, dx}{e^x - 1} \qquad (3.49)$$

After extending the upper limit to infinity instead of cutting off at x_m, and using the tabulated value of the pertinent Riemann zeta function (for $s = 3$), we have

$$\int_0^\infty \frac{x^2\, dx}{e^x - 1} = 1.202 \times 2 = 2.404 \qquad (3.50)$$

Thus, substitution of Eq. (3.50) into (3.49) yields

$$E_{T \to 0} \approx \frac{14.4 Nk_B}{\Theta_D^2} T^3 \qquad (3.51)$$

and the surface heat capacity is given by

$$C_V = \left(\frac{\partial E}{\partial T}\right)_V = \frac{43.2 Nk_B}{\Theta_D^2} T^2 \qquad (3.52)$$

We see that at low temperatures the surface heat capacity C_V is proportional to T^2, as opposed to the T^3 dependence of the bulk lattice heat capacity. However, the model we have considered here, consisting of a surface layer of atomic thickness, is quite unrealistic; it would be difficult to measure the heat capacity of a single atomic layer. In most cases the solid samples that can be used in experiments are small particles of variable surface/volume ratio or thin films many atomic layers thick. It would therefore be important to consider the heat capacity of such a sample and to see what contribution, if any, the surface makes to the total vibrational heat capacity.

For a solid consisting of atoms in the bulk and on the surface the total number of vibrational modes, $3N$, can be expressed after Montroll[9a] as

$$3N = \frac{4\pi V}{3C_3} \nu_D^3 + \frac{\pi \mathscr{A}}{C_2} \nu_D^2 \tag{3.53}$$

where the first term on the right-hand side is the bulk contribution and the second term is the surface contribution, by use of the elastic continuum model. Solving the cubic equation,[9a] one obtains for the maximum frequency ν_D,

$$\nu_D = \left(\frac{9NC_3}{4\pi V}\right)^{1/3}\left[1 - \frac{\pi \mathscr{A}}{36C_2 N^{1/3}}\left(\frac{9C_3}{4\pi V}\right)^{2/3} + 0(N^{-2/3})\right] \tag{3.54}$$

where $0(N^{-2/3})$ indicates that all other terms proportional to $N^{-2/3}$ were neglected. Now we proceed as before, using the frequency distribution

$$g(\nu) = \frac{4\pi V}{C_3} \nu^2 + \frac{2\pi \mathscr{A}}{C_2} \nu \tag{3.55}$$

The integration is somewhat more complex than before, since it involves several terms. Also, the method of counting surface modes depends on the boundary conditions, i.e., the shape of the thin sample being considered. However, the boundary conditions only change the constants but do not affect the temperature dependence of the different terms. Derivations using different boundary conditions can be found in the literature.[9b,c] Both surface and bulk heat capacities have the same high-temperature limit, $3R$. We shall only give here the heat capacity obtained at low temperature for a rectangular solid:

$$\frac{C_V(T \to 0°\text{K})}{Nk_B} \approx 234\left(\frac{T}{\Theta_{\text{bulk}}}\right)^3 + 50\left(\frac{T}{\Theta_{\text{surface}}}\right)^2 N^{-1/3} \tag{3.56}$$

We can see the familiar T^3 and T^2 dependences of the bulk and surface terms, respectively.

It should be noted that the effective Debye temperature, which is characteristic of the vibration of surface atoms, Θ_{surface}, may be different from Θ_{bulk}, which characterizes the vibration of bulk atoms. The experimental technique used to obtain Θ_{surface} will be discussed in the next section.

We would like to find out what is the contribution of the surface heat capacity to the total lattice heat capacity at a given temperature for a sample of given size. Since the surface heat capacity is proportional to the surface area and the bulk term is proportional to the volume, the surface/volume ratio will clearly play an important role in determining the magnitude of the contribution of the surface heat capacity to the total heat capacity. The ratio of the bulk and surface-heat-capacity terms, γ, indicates temperature range and the thickness of the specimen for which the surface-heat-capacity contribution will become detectable. The ratio γ for a cube with sides of length l is approximately given by

$$\gamma \approx 3.6 \times 10^8 l \frac{\Theta_{\text{surface}}^2}{\Theta_{\text{bulk}}^2} \left(\frac{\rho}{M}\right)^{1/3} T \tag{3.57}$$

where ρ is the density and M is the atomic weight of the sample. For a thin film of volume $l \times l \times ql$, where $ql \ll l$, or for a wire of length l and diameter ql, where again $ql \ll l$, we have

$$\gamma \approx 11 \times 10^8 ql \frac{\Theta_{\text{surface}}^2}{\Theta_{\text{bulk}}^3} \left(\frac{\rho}{M}\right)^{1/3} T \tag{3.58}$$

For most metals $(\rho/M)^{1/3} \approx 0.5 \text{ cm}^{-1}$; for platinum, for example, $\Theta_{\text{bulk}} = 234°\text{K}$ and $\Theta_{\text{surface}} \approx 110°\text{K}$; $\gamma \approx 5$ for a 1000-Å-thick film. Thus samples in the 10^{-5}–10^{-4}-cm thickness range should show a detectable contribution of surface heat capacity at temperatures $T \leqslant 10°\text{K}$.

There have been several attempts to measure surface heat capacity on samples of large surface/volume ratio. The T^2 dependence has been reported for the low-temperature heat capacity of graphite.[10] This material, having a layer structure, appears to have two-dimensional vibrational modes which contribute dominantly to the heat capacity in a certain temperature range. Surface-heat-capacity measurements using MgO and NaCl powders have also been reported.[11] However, the T^2 temperature dependence has not been established yet for these materials.

3.3. Mean-square Displacement of Surface Atoms

The *average* potential energy of a classical harmonic oscillator is $\frac{1}{2}kT$. Thus, from Eq. (3.4) we have

$$\tfrac{1}{2}K\langle x^2\rangle = \tfrac{1}{2}m\omega^2\langle x^2\rangle = \tfrac{1}{2}k_B T \tag{3.59}$$

where $\langle x^2 \rangle$ is the mean-square displacement. Since the average energy $\bar{E}$ of a harmonic oscillator is twice the average potential energy,[4] (see Eq. 3.13), we can express the mean-square displacement after rearrangement of Eq. (3.59), as

$$\langle x^2 \rangle = \frac{\bar{E}}{m\omega^2} \qquad (3.60)$$

By using the high-temperature limit for the total energy of N harmonic oscillators ($\bar{E} = 3Nk_BT$) and equating ω with the Debye cutoff frequency, $\omega_D = k_B\Theta_D/\hbar$, where $\omega_D = 2\pi\nu_D$ and $\hbar = h/2\pi$, Eq. (3.60) can be rewritten as

$$\langle x^2 \rangle = \frac{3N\hbar^2}{mk_B\Theta_D^2} T \qquad (3.61)$$

Thus the mean-square displacement of atoms that vibrate as harmonic oscillators is linearly proportional to the temperature at high temperatures ($T > \Theta_D$).

It has been found by experiment that the intensity of diffracted low-energy electrons or x-ray beams markedly depends on the temperature of the scattering solid. The intensities of the diffracted beams decrease exponentially with increasing temperature. This is because of the fact that at any one instant many of the vibrating atoms are displaced from their equilibrium position. Thus the incident electron beam encounters a partially disordered lattice. The atoms that are displaced from their equilibrium position during the scattering process will scatter out of phase and a fraction of the elastically scattered electrons will be found in the background instead of in the diffraction spot. The larger the amplitude of vibration, the more likely that the back-scattered electrons will be in the background instead of contributing to the diffraction spot intensity. It can be shown[12,13] that the intensity of a diffracted beam (neglecting multiple scattering events) is given by

$$I_{hkl} = |F_{hkl}|^2 \exp\left\{-[16\pi^2 (\cos^2 \phi)/\lambda^2]\langle x^2 \rangle\right\} \qquad (3.62)$$

where the exponential term is the *Debye–Waller factor*, λ is the electron wavelength, ϕ is the angle of incidence with respect to the surface normal, and $|F_{hkl}|^2$ is the scattered intensity caused by the rigid lattice. The constant terms $16\pi^2 (\cos^2 \phi)/\lambda^2$ gives the magnitude of the scattering vector, $|\Delta\mathbf{k}| = |\mathbf{k} - \mathbf{k}_0| = (4\pi/\lambda) \cos \phi$, squared, and $\langle x^2 \rangle$ is the mean-square displacement component in the direction of $\Delta\mathbf{k}$. Substituting

Eq. (3.61) into (3.62), we have

$$I_{hkl} = |F_{hkl}|^2 \exp\left[-\left(\frac{12Nh^2}{mk}\right)\left(\frac{\cos\phi}{\lambda}\right)^2 \frac{T}{(\Theta_D)^2}\right] \tag{3.63}$$

Thus according to Eq. (3.63) the diffracted beam intensity decreases exponentially with increasing temperature. By measuring as a function of temperature, the intensity of low-energy electrons diffracted from surface atoms, the mean-square displacement of surface atoms can be obtained. According to Eq. (3.63) the logarithm of the intensity plotted as a function of temperature T gives a straight line. The root-mean-square (rms) displacement of surface atoms can be calculated from the slope, the surface Debye temperature, and by using Eq. (3.62).

The rms displacement of surface atoms in several cubic metals has been determined in this manner. In most cases the mean displacement component, which is perpendicular to the surface plane, $\langle x_\perp^2 \rangle^{1/2}$, was obtained from low-energy electron diffraction experiments. The surface/bulk rms displacement ratios and the surface and bulk Debye temperatures are listed in Table 3.1. It can be seen that the rms displacement of surface

Table 3.1. *Surface and Bulk Root-mean-square Displacement Ratios and Debye Temperatures for Several Metals*

	$\dfrac{\langle u_\perp^2 \rangle^{1/2}\text{(surface)}}{\langle u^2 \rangle^{1/2}\text{(bulk)}}$	Θ(surface) (°K)	Θ(bulk) (°K)
Pb(110), (111)[29,30]	2.43 (1.84)	37 (49)	90
Bi(0001), (01$\bar{1}$2)[30]	2.42	48	116
Pd(100), (111)[29]	1.95	142	273
Ag(100),[31] (110),[31] (111)[32]	2.16 (1.48)	104 (152)	225
Pt(100), (110), (111)[33]	2.12	110	234
Ni(110)[34]	1.77	220	390
Ir(100)[35]	1.63	175	285
Cr(110)[36]	1.34	333	600
Nb(110)[37]	1.63	106	281

atoms, perpendicular to the surface plane, is 1.4 to 2 times as large as the bulk value. Similar large rms displacements were obtained for different crystal faces of the same solid in most cases. Correspondingly, the calculated surface Debye temperatures are smaller than their bulk values.

Debye–Waller measurements of this type do not yield the "net" displacement of surface atoms. The larger mean-square vibrational amplitude indicates, however, that the surface atoms either displace outward to new equilibrium positions by a 2–5 per cent increase of the interplanar distance and maintain nearly harmonic motion or they remain in their equilibrium positions similar to those for bulk atoms and their vibration becomes strongly anharmonic. Although there should be marked anharmonicity associated with the vibration of surface atoms at high temperatures, the large mean-square displacement values point to the presence of a net outward displacement of the surface layer of similar magnitude (a few per cent of the interplanar distance) as calculated theoretically for the different crystal surfaces.[14]

The mean-square displacement of surface atoms should be sensitive to changes in the number and type of neighboring atoms that can strongly interact with the surface. The adsorption on the clean surface of gases that chemically interact with the surface atoms (for example, oxygen on nickel or tungsten) should strongly affect the vibrational amplitude of surface atoms. This effect, which has been detected for tungsten surfaces in the presence of oxygen,[15] can be used to obtain valuable information about the thermal motion of surface atoms in different chemical environments.

3.4. Surface Modes of Lattice Vibrations

Since surface atoms have fewer neighbors than bulk atoms, there is a reduction in the interatomic forces at the surface when compared with the bulk. We have seen many consequences of this already. For example, the mean-square amplitude of atomic vibrations at the surface increased with a corresponding decrease in the vibrational frequencies. Another consequence of the creation of the surface is that certain modes of vibrations become localized; i.e., these vibrations only propagate along the surface and their displacement dies out exponentially toward the interior of the crystal.

Localized surface lattice vibrations were first predicted by Lord Rayleigh,[16] who studied the properties of the elastic continuum. He found that the equations of motion governing the propagation of lattice waves in such a medium have solutions that give waves that are freely propagating along the surface. These surface modes of lattice vibrations are frequently called *Rayleigh waves*. The velocity of these surface waves is always somewhat smaller than the velocity of the bulk lattice waves. More recently, several types of vibrational modes with different frequencies have been found, by careful analysis of the equations of motion in the elastic continuum under different boundary conditions, which are all restricted to the surface.[17,18] The effect of introducing two types of atoms with different masses on the distribution of surface modes of lattice vibrations has also been investigated.[19] The effect of impurities on these surface modes has been considered in detail as well.[20]

There have been few experimental studies aimed at determining the surface modes of lattice vibration. Electron scattering studies, in which the energy losses (millivolts) of low-energy electrons (few volts) that are due to the excitation of surface phonons were monitored, were reported. It is likely that atomic-beam-scattering experiments, in which the energy losses and the angular distribution of the scattered atoms are analyzed, will provide a new experimental technique for these studies. This important field of surface science still awaits innovations in experimental methods of measurement.

3.5. Surface Diffusion

3.5.1. INTRODUCTION

In this chapter we have so far discussed several phenomena associated with the vibration of surface atoms about their equilibrium positions. The thermal energy that excites atoms and forces them to oscillate with increasing amplitudes as the temperature is increased, however, is not sufficient to dislodge most of them from their equilibrium positions. The thermal energy ($3RT \approx 1.8 \text{ kcal/mole}$ at $300°K$) tied up in lattice vibrations is only a small fraction of the total energy necessary to break away an atom from its neighbors and to move it along the surface. This energy, as we shall see below, is on the order of 15–50 kcal/mole for many metal surfaces. Nonetheless, as the temperature of the surface is increased, more and more surface atoms may acquire enough activation energy to break bonds with their neighbors and move along the surface. Such surface diffusion plays an important role in many surface phenomena involving atomic transport, e.g., crystal growth, vaporization, and adsorption. Therefore, it is essential that we discuss the principles of surface diffusion and then show some of the results of recent investigations.

Figure 3.5 gives a schematic representation of a heterogeneous surface. Experimental evidence indicates that one can distinguish several atomic positions that differ in the number of neighbors surrounding them. Atoms in different surface sites have different binding energies. Surface diffusion is considered to be a multistep process in which atoms break away from their lattice position (e.g., a kink site at a ledge) and migrate along the surface until they find their new equilibrium site.

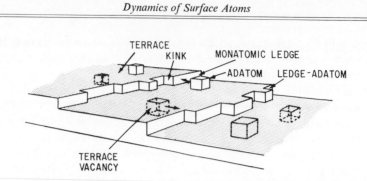

Figure 3.5. *Model of the heterogeneous surface depicting the different surface sites: terrace, kink, ledge, vacancy, ledge–ad-atom, and ad-atom.*

Let us assume that the only diffusing species are ad-atoms (adsorbed atoms) held on the surface by a small binding energy with respect to surface atoms in other lattice positions. In order to move an ad-atom to a neighboring site, a certain amount of thermal energy is needed. Since the atom can only occupy equilibrium sites at the beginning and at the end of the jump on an ordered crystal surface, the atom in the region in between the two sites must be in a higher energy state. The energy difference between the ad-atom at this energy maximum and that at an equilibrium lattice site is designated ΔE_D. During its thermal vibration about the equilibrium site, the atom strikes against this potential-energy barrier ν_0 times per second. Most of the time its energy is inadequate to pass over the energy barrier, but occasionally energy fluctuations increase its energy to ΔE_D^* and it crosses the barrier to its new equilibrium position. Thus the jump frequency to a neighboring site is given by $\nu_0 \exp(-\Delta E_D^*/k_B T)$. Since the ad-atom can jump into $\mathscr{y}$ equivalent neighboring sites, the total jump frequency f is given by

$$f = \mathscr{y}\nu_0 \exp\left(\frac{-\Delta E_D^*}{k_B T}\right) \tag{3.64}$$

For a (111) face of a face-centered cubic metal, $\mathscr{y} = 6$; the vibrational frequency is on the order of 10^{12} sec^{-1}. Assuming that ΔE_D is 20 kcal/ mole, at 300°K the atom makes one jump in every 50 sec and at 1000°K one in 10^{-8} sec. Thus we can readily see that the surface diffusion rate changes very rapidly with temperature.

Let us now consider that the ad-atom concentration in the surface is very small. Since we have assumed that surface diffusion can only occur via the motion of ad-atoms, these species have to be created before any appreciable diffusion can occur. The fraction of ad-atoms, c/c_0, in the surface is given by $c/c_0 = \exp(-\Delta E_f/k_B T)$, where ΔE_f is the energy of formation of an ad-atom, c is the concentration of ad-atoms, and c_0 is the total concentration of surface sites from which the ad-atom can break away. Now, the total jump frequency is further reduced by another exponential factor:

$$f = \nu_0 \mathcal{Y} \exp \left[- \frac{(\Delta E_D^* + \Delta E_f)}{k_B T} \right] \qquad (3.65)$$

For many cubic metals ΔE_f is about 40 kcal/mole. Thus if the ad-atom has to be created before it can jump along the surface, the atom makes one jump every 10^{31} sec (i.e., 10^{24} years) at 300°K and about one jump in every 10^{-4} sec at 1000°K. So we see that any change in the mechanism of diffusion that affects the overall activation energy for the process can markedly change the jump frequency.

3.5.2. RANDOM WALK

So far we have considered the conditions necessary for an atom to make a single jump to a neighboring equilibrium surface site. We would now like to derive an equation for the long-distance motion of a given surface atom. One would like to know, starting at a certain surface site, where the diffusing atom will be on the surface after time t and after a very large number of jumps. In order to calculate the net displacement of an atom we assume that the atomic motion is completely random. However, the jumps are of equal length, which is equal to the nearest-neighbor distance d. The derivation is the simplest if we consider the *random walk* of diffusing atoms only in one dimension;[21] in Fig. 3.6 we show the

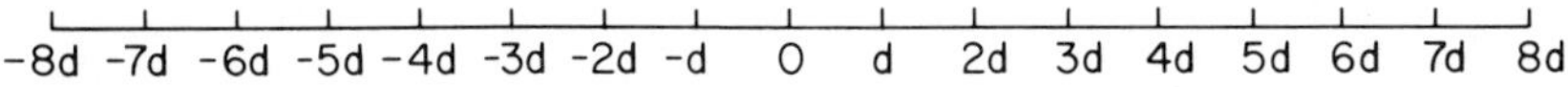

Figure 3.6. *Coordinates for computing the mean-square displacement for the one-dimensional random walk.*

coordinates for the one-dimensional motion. The net distance X covered by the atom is, of course, equal to the algebraic sum of the individual jumps in the range $-nd$ to $+nd$, where n is the number of jumps. The *average* distance traveled by the atom, $\langle X \rangle$, which is the algebraic average of X, is zero since jumps in both the positive and negative directions are equally probable. There are other averages (nonalgebraic), such as the rms distance, that are not zero and allow us to compute the total distance traveled by the atom from the starting point. The square of the net distance, X^2, is given by

$$X^2 = (d_1 + d_2 + \cdots + d_n)(d_1 + d_2 + \cdots + d_n) = d_1^2 + d_2^2 + \cdots$$
$$+ d_n^2 + 2d_1d_2 + 2d_1d_3 + \cdots + 2d_1d_n + 2d_2d_3 + 2d_{n-1}d_n \cdots \quad (3.66)$$

The average value of X^2, which we call the mean-square distance and designate $\langle X^2 \rangle$, is given by the sum of the averages of the individual terms in Eq. (3.66). Each of the squared terms is equal to d^2 since their magnitudes are equal; $|d_1| = |d_2| = \cdots = |d_n| = d$. The cross terms, $\cdots 2d_{n-1}d_n$, are equal to zero when averaged, since each of the $d_1 \cdots d_n$ terms has an equal chance to be positive or negative. Therefore, the mean-square distance is equal to

$$\langle X^2 \rangle = d_1^2 + d_2^2 + \cdots + d_n^2 = nd^2 \quad (3.67)$$

The number of jumps n can be expressed as the product of the total jump frequency f and the time t required to make n number of jumps $n = ft$. Hence the mean-square distance in one dimension is given by

$$\langle X^2 \rangle = ftd^2 \quad (3.68)$$

Inspection of some of our previous equations [(3.65) for example] indicates that fd^2 is the property of a given material. If atomic jumps can occur with equal probability along three different coordinates instead of only one, as on a (111) face of a cubic crystal that has sixfold rotational symmetry, the mean-square displacement in one direction is reduced to†

$$\langle X^2 \rangle = \frac{ftd^2}{3} \quad (3.69)$$

If the product of the jump frequency and the interatomic distance, fd^2, is a property of a material that characterizes its atomic transport it is worthwhile to determine its magnitude for as many materials as possible. Its value should provide information about the mechanism of atomic transport. Thus it is customary to define the *diffusion coefficient* D as

$$D = \frac{fd^2}{2b} \quad (3.70)$$

† For a square surface unit mesh with fourfold rotational symmetry, the mean-square displacement is given by $\langle X^2 \rangle = ftd^2/2$.

where b is the number of coordinate directions in which diffusion jumps may occur with equal probability. For diffusion that is equally probable in two directions we can write the diffusion coefficient after substitution of Eq. (3.65) into (3.70) as

$$D = \frac{d^2 \nu_0}{4} \exp\left[-\frac{(\Delta E_D^* + \Delta E_f)}{k_B T}\right] \tag{3.71}$$

Thus D shows exponential temperature dependence, which was confirmed experimentally for most solids. D, in general, is given in units of square centimeters per second. If we can determine D by experiments, from the straight slope of the ln D versus $1/T$ plot, the activation energy of the diffusion process can be determined, provided that diffusion occurs by a single mechanism. If there is a change from one diffusion mechanism in the studied temperature range to another mechanism with a different activation energy, the slope of the line of the ln D versus $1/T$ plot will change.

The rms distance $\langle X^2 \rangle^{1/2}$ can be expressed in terms of the diffusion coefficient by substitution of Eq. (3.70) into (3.69) to yield for $b = 1$,

$$\langle X^2 \rangle^{1/2} = (2Dt)^{1/2} \tag{3.72}$$

Thus measurement of the mean travel distance of diffusing atoms allows one to evaluate the diffusion coefficient. Conversely, knowledge of the diffusion coefficient allows one to estimate the rms distance or the time necessary to carry out the diffusion. For example, the diffusion coefficients of silver ions on the surface of silver bromide can be estimated to be 10^{-9} and 10^{-13} cm²/sec at 300 and 100°K, respectively. Assuming that a rms distance of 10^{-4} cm is required for silver particle aggregation (print-out) to commence, of what duration are the light-exposure times required? Using Eq. (3.72) we have $t = 5$ and $t = 5 \times 10^4$ sec at 300 and 100°K, respectively. The exponential temperature dependence of D is, of course, the reason that silver bromide photography cannot be carried out at low temperatures (much below 300°K) but is easily utilized at about room temperature. We can also see that at slightly elevated temperatures ($\sim 450°$K) the thermal diffusion of silver particles should be rapid enough ($D \approx 3 \times 10^{-7}$ cm²/sec) so that their aggregation will take place rapidly even in the dark ($t \approx 10^{-2}$ sec) in the absence of any photoreaction.

3.5.3. MACROSCOPIC DIFFUSION PARAMETERS

So far we have described the diffusion of a single surface atom, i.e., its motion across a potential-energy barrier that reduces its jump frequency and its distance of travel. However, on a real surface many atoms diffuse simultaneously and their concentration could be in the range 10^{10}–10^{13} atoms/cm². Therefore, in diffusion experiments the measured diffusion distance after a given diffusion time is an average of the diffusion lengths of a large, statistical number of surface atoms. It is thus important to define the diffusion process in terms of macroscopic parameters. We are going to follow a treatment similar to that by Wert and Zener.[22]

Let us assume that among ad-atoms in different energy states, there is equilibrium characterized by the Boltzmann distribution. The jump frequency [Eq. (3.64)] characteristic of all the diffusing atoms can then be expressed as

$$f = \gamma v_0 \exp\left(-\frac{\Delta G_D^*}{RT}\right) \tag{3.73}$$

where ΔG_D^* is the increase in free energy in moving 1 mole of particles from their equilibrium energy state to their state at the top of the potential barrier. We can rewrite Eq. (3.73) as

$$f = \gamma v_0 \exp\left(\frac{\Delta S_D^*}{R}\right) \exp\left(-\frac{\Delta H_D^*}{RT}\right) \tag{3.74}$$

where ΔH_D^* and ΔS_D^* are the activation energy (or enthalpy) and activation entropy for the diffusion process.† If the diffusion mechanism requires the formation of the diffusing species as well, we have to take into account the free energy of formation of 1 mole of ad-atoms (or other diffusing *entities*); therefore, Eq. (3.74) becomes[23]

$$f = \gamma v_0 \exp\left[-\frac{(\Delta G_D^* + \Delta G_f)}{RT}\right] \tag{3.75}$$

or

$$f = \gamma v_0 \exp\left(\frac{\Delta S_D^* + \Delta S_f}{R}\right) \exp\left[-\frac{(\Delta H_D^* + \Delta H_f)}{RT}\right] \tag{3.76}$$

The diffusion coefficient D is now averaged over all of the diffusing atoms and Eq. (3.71) becomes

$$D = \gamma \frac{d^2 v_0}{4} \exp\left(\frac{\Delta S_D^* + \Delta S_f}{R}\right) \exp\left[-\frac{(\Delta H_D^* + \Delta H_f)}{RT}\right] \tag{3.77}$$

† We can use the terms *activation energy* and *enthalpy* for ΔH_D^* interchangeably since the difference, $\Delta(PV)$, is very small for a solid state or surface diffusion process and therefore it can be neglected.

By grouping the temperature-independent terms in Eq. (3.77) into one constant, D_0, which is called the diffusion constant, and letting $Q = \Delta H_D^* + \Delta H_f$, where Q is often called the total activation energy for the overall diffusion process, we have

$$D = D_0 \exp\left(\frac{-Q}{RT}\right) \tag{3.78}$$

Since $\nu_0 \approx 10^{12}\ \text{sec}^{-1}$ and $d \approx 2 \times 10^{-8}$ cm, D_0 has the minimum value of $D_0 \approx 6 \times 10^{-4}\ \text{cm}^2/\text{sec}$ (assuming that the ΔS_D^* and ΔS_f are zero). However, experimental values of D_0 range widely between 10^{-3} and $10^3\ \text{cm}^2/\text{sec}$ for the self-diffusion of atoms in most solid surfaces, as will be shown shortly. This indicates that the entropy change associated with the diffusion process is always positive. Similarly, Eq. (3.72) will also reflect the rms travel distance of a statistical number of diffusing surface atoms. In practice, most surface diffusion experiments will measure the diffusion characteristics of a large concentration of atoms since most of the detection techniques (except the technique of field-ion microscopy) are not yet sensitive enough to monitor the motion of single atoms.

Now that we have defined the terms used to describe the diffusion of atoms and discussed some of the principles of diffusion, let us calculate the rate of diffusion in one dimension. Consider three parallel atomic rows on the surface: A, B, and C. Ad-atoms, which will still be considered to be the major diffusing entities, may jump from rows A and C to B. Thus there is a flow of atoms to B from both directions. Under certain circumstances, there exists a net flux in one direction; Let us only consider the flow of atoms from A to B.

We define the concentration of ad-atoms in rows A and B as c_A and c_B. We assign a length l to every atomic row and a "thickness" that is equal to the interrow distance d (Fig. 3.7). If we have N number of

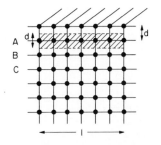

Figure 3.7. *Model to derive the rate of diffusion in one dimension.*

atoms in an area ld occupied by the atomic row A, the atomic concentration in that row is $c_A = N_A/ld$. Since f is the number of jumps per second, the net flow to B, dN_B/dt (number of atoms per second), is given by

$$\frac{dN_B}{dt} = \frac{1}{2} f(N_A - N_B) = \frac{1}{2} fld(c_A - c_B) \qquad (3.79)$$

where the $\frac{1}{2}$ indicates that in each row the ad-atoms have equal chance to jump toward B or away from it, hence that the total concentrations available to move toward B is $\frac{1}{2}c_A$. If $c_A = c_B$, the net flow into B from one direction is zero. However, in most diffusion experiments we can establish a concentration gradient; i.e., we can establish conditions where $c_A \neq c_B$. If there is a net flow of atoms with $c_C < c_A$, for example, it

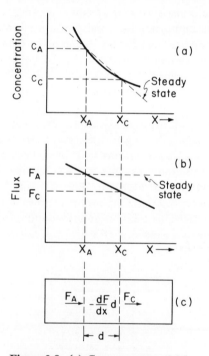

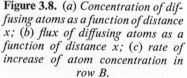

Figure 3.8. (*a*) *Concentration of diffusing atoms as a function of distance* x; (*b*) *flux of diffusing atoms as a function of distance* x; (*c*) *rate of increase of atom concentration in row B.*

can be seen from Fig. 3.8(a) that $c_A - c_C$ can be expressed as

$$c_A - c_C = -\frac{dc}{dx} d \tag{3.80}$$

where x is the distance along the direction of diffusion. Substituting Eq. (3.80) into (3.79) and remembering (3.70) we have the flux

$$\frac{1}{l}\frac{dN_B}{dt} = -\tfrac{1}{2}fd^2\frac{dc}{dx} = -D\frac{dc}{dx} \tag{3.81}$$

Thus the net number of atoms entering atomic row B from the adjacent rows per second per unit length is given by the product of the diffusion coefficient and the concentration gradient. Equation (3.81) is similar to Fick's law, which was derived for three-dimensional diffusion. For bulk diffusion the volume concentration is defined as $c_V = N/l^2d$ and the flux has units of atoms per square centimeter per second.

The minus sign in Eq. (3.81) indicates that the flux is always in the direction of decreasing concentration. If the flux of atoms through B is constant (does not vary with time), we have

$$-D\frac{dc}{dx} = \text{constant} \tag{3.82}$$

Under these conditions the concentration of ad-atoms, c_B, remains constant throughout the diffusion process. By plotting the concentration as a function of distance along the surface we obtain a straight line. This is the condition for *steady-state* diffusion. If one establishes steady-state conditions in surface diffusion experiments, the slope of the c versus x plot will give the diffusion coefficient directly.

Frequently, steady-state diffusion conditions cannot be established, and thus the flux of ad-atoms in row B will change as a function of time [Fig. 3.8(b)]. This indicates a time-dependent accumulation or depletion of the ad-atom concentration c_B. Let us consider again diffusion in one dimension and assume that the ad-atoms actually accumulate in B. The flux of atoms into B is given by $F - (dF/dx)d$, where $F = (1/l)(dN/dt)$ and $-(dF/dx)d$ gives the change of flux in the element occupied by the atomic row B. The negative sign again indicates that the change of flux occurs in the direction of decreasing concentration. The flux of atoms out of B is equal to F. Thus the rate at which the concentration of the diffusing substance increases in *B* is given by [Fig. 3.8(c)]

$$\frac{\partial C_B}{\partial t} d = \left(F - \frac{\partial F}{\partial x} d \right) - F \tag{3.83}$$

Substitution for *F* from Eq. (3.81) yields

$$\frac{\partial c}{\partial t} = D \frac{\partial^2 c}{\partial x^2} \tag{3.84}$$

which is often referred to as *Fick's second law* of diffusion in one dimension.

3.5.4. SURFACE DIFFUSION STUDIES

In most surface diffusion studies the experimental data (most frequently measured is the surface concentration of diffusants *c* as a function of distance *X* along the surface) are analyzed by solving Eq. (3.84) by the use of boundary conditions that approximate well the experimental geometry. For example, let us consider the diffusion of ad-atoms from one end of the surface to the other. These surface atoms are "tagged" by using a radioactive tracer such as a Ni isotope whose diffusion is monitored on a nickel (111) single-crystal surface by a suitable detector.[24a] If the surface concentration remains constant throughout the diffusion at one point on the surface where diffusion is initiated and the concentration of ad-atoms is initially zero everywhere else on the surface the solution of Eq. (3.84) is of the form[24b]

$$c(x, t) \approx \frac{c_0 e^{-\alpha x}}{x^{1/2}} \tag{3.85}$$

where α is given by[24]

$$\alpha^2 = \frac{2}{D\delta} \left[\frac{D_V}{(\pi D_V t)^{1/2}} + \frac{\lambda_1 \nu_0 \delta}{2} e^{-(5/6)\Delta H_s/RT} \right] \tag{3.86}$$

D and D_V are the surface and bulk diffusion coefficients, respectively; λ_1 is the average number of interatomic distances between ledges on the surface; δ is the surface-layer thickness (in this case $\delta = d$); and ΔH_s is the heat of sublimation. Since the activity of the radiotracer, *A*, as

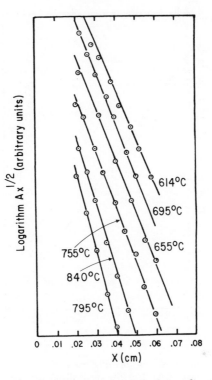

Figure 3.9. *Logarithm of the radio-tracer activity as a function of diffusion distance for a nickel isotope on the (111) crystal face of nickel.*

measured by the detector (e.g., autoradiographic plate) is proportional to the surface concentration $c(x, t)$, a plot of the logarithm of $Ax^{1/2}$ versus x after a given surface diffusion time at a given temperature should be linear with a slope equal to α. The log $Ax^{1/2}$ versus x plots are shown in Fig. 3.9 for different surface temperatures. The surface diffusion coefficients $D(T)$, which are calculated from the slopes, are given in the log D versus $1/T$ plot in Fig. 3.10. The activation energy for surface self-diffusion of nickel is $\Delta H_D^* = 38 \pm 4$ kcal/mole [$D(\text{cm}^2/\text{sec}) = 300 \exp(-38/RT)$] using the following values for the parameters that appear in Eq. (3.86):

$$D_v\left(\frac{\text{cm}^2}{\text{sec}}\right) = 1.27 \exp\left[\frac{-66.9(\text{kcal/mole})}{RT}\right]$$

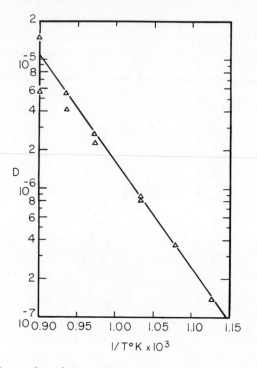

Figure 3.10. *Logarithm of the surface diffusion coefficient as a function of the reciprocal temperature for the surface self-diffusion of nickel.*

$\Delta H_s = 100$ kcal/mole, $\delta = 2.5 \times 10^{-8}$ cm, and $\lambda_1 = 5$. The measured D and D_0 values are consistent with a single diffusion mechanism throughout the studied temperature range (600–800°C) in which adatoms move easily along the atomic terraces on the single-crystal surface but are trapped at ledges that are, on the average, spaced $5d$ apart.

In most of the diffusion studies it is assumed that D, the diffusion coefficient, is constant and independent of the concentration of the diffusing species. Although in most cases this assumption is valid, there are certain diffusion processes where D may not be constant. There are excellent books[25,26] that give detailed analyses of the various solutions of Fick's laws under different boundary conditions and their modification if D is a function of the concentration. The reader should consult these references when working on diffusion problems.

Surface diffusion studies on solid surfaces have been carried out by using several experimental techniques. Among these, radioactive tracer diffusion and grain-boundary-grooving techniques appear to be those most frequently employed.[27] We have already briefly described the radiotracer technique above. The latter method involves the introduction of a well-defined groove by scratching or chemically etching the surface; then the change of the groove slope is followed by interference microscopy or another suitable technique. The change of groove slope results from the mass transport (surface diffusion) of atoms and thus the diffusion rate at a given temperature can be determined.

The difficulty in most surface diffusion experiments is to determine the type of surface species moving along the surface. The overall diffusion rate can be determined with relative ease and the diffusion coefficient can be calculated. However, D obtained in this way does not give one direct information about the diffusing entities. The heterogeneous surface contains atoms in different lattice positions in which their binding energies are different (see Fig. 3.5). The surface diffusion measurements do' not easily distinguish among the atoms to indicate which of the surface species contributes dominantly to the diffusion flux. The temperature dependence of D will give the overall activation energy of diffusion, Q. Its magnitude often enables one to distinguish between different overall mechanisms of surface transport. For example, if ad-atoms, already present on the surface in large concentrations, are the dominant contributors to the diffusion flux, the activation energy will be $Q = \Delta H_D^*$. If the ad-atom concentration is small and the diffusion process includes their formation and subsequent diffusion, the activation energy will be greater and be given by $Q = \Delta H_D^* + \Delta H_f$. If vacancies at the surface are the majority mobile species, the activation energy may be equal to their energy of formation plus the activation energy of their subsequent diffusion.

It is often observed that one type of surface diffusion technique gives reproducible D and Q values that are, however, different from the D and Q values obtained by a different experimental technique for the same system. It appears that the surface preparation used in applying the technique establishes the surface condition (the concentration and type of diffusing entities), which is different from surface conditions established by other techniques. Often the thermal history of the samples and the ambient atmosphere affect the results of surface diffusion studies. Investigations by field-ion microscopy (to be discussed in Chapter 4) appear to be able to distinguish and identify the different diffusing species (ad-atoms, vacancies, atoms in ledge, etc.) and hold the hope that the activation energies of diffusion of each of these entities may be determined separately for a given surface.

In Table 3.2 the diffusion constant and the overall activation energies of surface diffusion Q are listed for several solids. These values were obtained from self-diffusion studies carried out by using different experimental techniques.

Since the formation of more ad-atoms or any other mobile species that have smaller binding energies increases the surface disorder, ΔS_f is positive, in general. We have shown that the minimum value of D_0 is on the order of 10^{-4} cm²/sec. Larger values indicate either large surface entropies—i.e., the diffusing molecule may have one or more translational degrees of freedom—or jump distances larger than the interatomic

Table 3.2. *Total Activation Energies for Surface Self-diffusion Q and Diffusion Constants D_0 for Several Metals*

Material	Q (kcal/mole)[28]	D_0 (cm²)/sec)
Ni	38.0	300
Pt	26.1–30	4×10^{-3}
Rh	41.5	$\sim 4 \times 10^{-2}$
Re	52.0	~ 1.0
W	68–78	0.85
Cu	41–46	~ 650
Au	35–42	0.37
Ta	45	—
Mo	52–56	~ 0.8
Fe	59.6	—

spacing. The activation energies of surface diffusion are markedly smaller than the corresponding values for bulk diffusion. Thus rates of surface diffusion are orders of magnitude larger at a given temperature than bulk diffusion rates.

Figure 3.11 shows a plot of the logarithm of the self-diffusion coefficients for surface diffusion for several face-centered cubic metals as a function of the "reduced" reciprocal temperature T_M/T, where T_M is the melting temperature.[28] As long as the diffusion mechanism is the same, the resulting slope should define a nearly straight line. There is, however, a definite curvature in the slope as a function of the temperature (with the exception of silver). It appears that the mechanism of surface diffusion of these face-centered cubic metals changes as a function of temperature. Figure 3.12 also shows the large scatter of the data due to the experimental uncertainties (probably impurity effects) mentioned above. It has been suggested that, at low temperature, the motion of ad-atoms dominates surface diffusion. This process appears to have an activation energy $Q = 0.24\Delta H_s$, where ΔH_s is the heat of sublimation of the solid. At high temperatures the dominant carriers of the surface diffusion flux

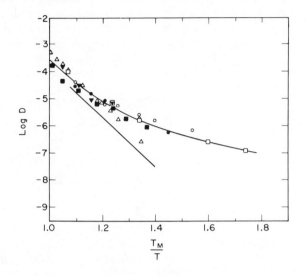

Figure 3.11. *Logarithm of the surface self-diffusion coefficients of several face-centered cubic metals as a function of the reduced reciprocal temperature* T_M/T. $\triangle$, $\square$: *gold*; $\blacksquare$, $\blacktriangledown$, $\bigcirc$: *copper*; $\bullet$: *nickel*; – –: *silver*.

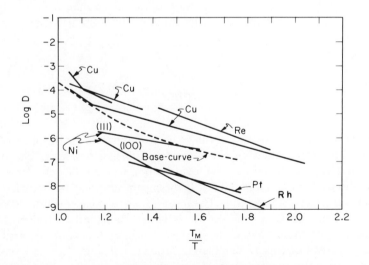

Figure 3.12. *Logarithm of the surface self-diffusion coefficients of various face-centered cubic metals as a function of the reduced temperature* T_M/T, *showing large scatter due to impurities.*

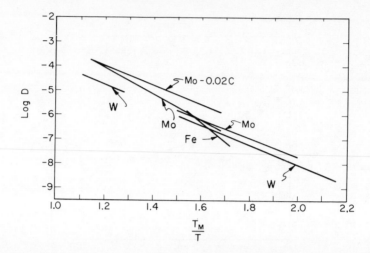

Figure 3.13. *Logarithm of the surface self-diffusion coefficients of several body-centered cubic metals as a function of the reduced reciprocal temperature T_M/T.*

are assumed to be ad-atom vacancy pairs with $Q = 0.54\Delta H_s$. A similar plot (log D versus $1/T$) for several body-centered cubic metals yields a straight line (Fig. 3.13). It appears that for surfaces of body-centered cubic solids there is only one surface diffusion mechanism for which $Q = 0.33\Delta H_s$ and which is operative throughout the studied temperature range.

Summary

The harmonic oscillator model can be used to approximate the properties of surface atoms vibrating about their equilibrium positions. The surface heat capacity makes a small contribution to the total heat capacity that should be detectable only at low temperatures and for samples of large surface to volume ratio. The mean square displacement of surface atoms perpendicular to the surface is larger than the bulk value. The migration of atoms along the surface (surface diffusion) requires a smaller activation energy than their diffusion in the bulk. The mechanism of surface diffusion, the type of surface species (ad-atoms, vacancies, atoms in ledge, etc.) that contribute dominantly to the diffusion flux may change as a function of temperature or other changing conditions (surface structure, ambient atmosphere, etc.) at the surface.

References

1. R. A. Levy: *Principles of Solid State Physics*. Academic Press, Inc., New York, 1968.

2. G. N. Lewis and M. Randall: *Thermodynamics*, revised by K. S. Pitzer and L. Brewer. (McGraw-Hill Book Company, New York, 1961.

3. A. Einstein: *Ann. Physik*, **22**, 180 (1907).

4. J. M. Ziman: *Principles of the Theory of Solids*. Cambridge University Press, New York, 1965.

5. P. Debye: *Ann. Physik.*, **39**, 789 (1912).

6. L. Brillouin: *Wave Propagation in Periodic Structures*. Dover Publications, Inc., New York, 1945.

7. T. L. Hill: *Introduction to Statistical Thermodynamics*. Addison-Wesley Publishing Company, Inc., Reading, Mass., 1962.

8. E. Jahnke and F. Emde: *Tables of Functions*. Dover Publications, Inc., New York, 1945.

9a. E. W. Montroll: *J. Chem. Phys.*, **18**, 183 (1950).

9b. M. Dupuis, R. Mazo, and L. Onsager: *J. Chem. Phys.*, **33**, 1452 (1960).

9c. R. Stratton: *Phil. Mag.*, **44**, 519 (1953).

10. J. Krumhansl and H. Brook: *J. Chem. Phys.*, **21**, 1663 (1953).

11a. W. F. Giauque and R. C. Archibald: *J. Am. Chem. Soc.*, **59**, 561 (1937).

11b. P. Balk and G. C. Benson: *J. Phys. Chem.*, **63**, 1009 (1959).

12. R. W. James: *The Optical Principles of the Diffraction of X-ray*. G. Bell & Sons Ltd., London, 1965.

13. B. C. Clark, R. Herman, and R. F. Wallis: *Phys. Rev.*, **139**, A860 (1965).

14. J. J. Burton and G. Jura: *J. Phys. Chem.*, **71**, 1937 (1967).

15. P. Estrup: in *The Structure and Chemistry of Solid Surfaces*, G. A. Somorjai, ed. John Wiley & Sons, Inc., New York, 1969

16. L. Raleigh: *Proc. London Math. Soc.*, **17**, 4 (1887).

17. D. C. Gazis, R. Herman, and R. F. Wallis: *Phys. Rev.*, **119**, 533 (1960).

18. D. C. Gazis and R. F. Wallis: *Surface Sci.*, **3**, 19 (1964).

19. R. F. Wallis: *Phys. Rev.*, **105**, 540 (1957).

20. H. Kaplan: *Phys. Rev.*, **125**, 1271 (1962).

21. C. A. Wert and R. M. Thompson: *Physics of Solids*. McGraw-Hill Book Company, New York, 1964.

22. C. A. Wert and C. Zener: *Phys. Rev.*, **76**, 1169 (1949).

23. P. G. Shewmon: *Diffusion in Solids*. McGraw-Hill Book Company, New York, 1963.

24a. J. R. Wolf and H. W. Weart: in *The Structure and Chemistry of Solid Surfaces*, G. A. Somorjai, ed. John Wiley & Sons, Inc., New York, 1969.

24b. P. G. Shewmon: *J. Appl. Phys.*, **34,** 755 (1963).

25. W. Jost: *Diffusion in Solids, Liquids and Gases.* Academic Press, Inc., New York, 1952.

26. H. S. Carslaw and J. C. Jaeger: *Conduction of Heat in Solids.* Oxford University Press, New York, 1959.

27. N. A. Gjostein: in *Fundamentals of Gas-Solid Interactions*, H. Saltsburg et al., eds. Academic Press, Inc., New York, 1967.

28. N. A. Gjostein: in *Surfaces and Interfaces*, J. J. Burke et al., eds. Syracuse University Press, Syracuse, N.Y., 1967.

29. R. M. Goodman, H. H. Farrell, and G. A. Somorjai: *J. Chem. Phys.*, **48,** 1046 (1968).

30. R. M. Goodman and G. A. Somorjai: *J. Chem. Phys.*, **52,** 6325 (1970).

31. J. M. Morabito, Jr., R. F. Steiger, and G. A. Somorjai: *Phys. Rev.*, **179,** 638 (1969).

32. E. R. Jones, J. T. McKinney, and M. B. Webb: *Phys. Rev.*, **151,** 476 (1966).

33. H. B. Lyon and G. A. Somorjai: *J. Chem. Phys.*, **44,** 3707 (1966).

34. A. U. McRae: *Surface Sci.*, **2,** 522 (1964).

35. R. M. Goodman: Ph.D. Dissertation, University of California, Berkeley, Calif., 1969.

36. R. Kaplan and G. A. Somorjai: *Solid State Comm.*, **9,** 505 (1971).

37. D. Tabor and J. Wilson: *Surface Sci.*, **20,** 203 (1970).

Problems

3.1. The lattice heat capacity of copper has been measured calorimetrically in the temperature range 10–300°K; the experimental data are on page 119. Plot the data and fit the heat-capacity equations that are based on both the Debye and the Einstein model to the data points; compute Θ_E and Θ_D.

3.2. Calculate the surface-heat-capacity contribution at 5°K to the total heat capacity of a silver thin film of dimensions $l = 1$ cm and $ql = 10^3$ Å.

3.3. The mean-square displacements of surface atoms perpendicular to the surface plane has been measured for lead and bismuth (see Table 3.1). In the same study the intensity of the (00) diffraction beam was monitored up to the melting point in order to monitor the temperature at which the surface disorders.[30] Melting of the surfaces has occurred at the bulk melting temperatures for both materials. (a) Review the evidence in support of premelting of surfaces of certain solids and explain the results in light of these evidences. (b) How would you expect various surface impurities (other metals, adsorbed oxygen, etc.) to influence the temperature at which the surface becomes automatically disordered? [Ref.: *J. Chem. Phys.*, **52,** 6325 (1970).]

3.4. A ^{187}W isotope was deposited on one end of a 5-cm-long tungsten slab. Estimate the time necessary to detect the radiotraces at the other end of the slab at 1000°C.

T (°K)	Copper heat capacity C (J/mole-°K)
10	0.05464
15	0.1716
20	0.4843
25	1.017
30	1.716
40	3.812
50	6.290
60	8.705
70	10.99
80	13.03
90	14.74
100	16.14
120	18.30
140	19.89
160	21.10
180	21.99
200	22.62
220	23.13
240	23.57
260	23.89
280	24.21
300	24.53

3.5. The surface self-diffusion of copper was measured by Collins et al., *Trans. AIME*, **236,** 1354 (1966), and Bradshaw et al., *Acta Met.*, **12,** 1057 (1964). Review these articles and suggest possible reasons for the discrepancy of their data.

3.6. There are several photosensitive compounds (AgBr, $PbCl_2$, PbI_2) that undergo thermal decomposition as well. Explosives that are known to undergo rapid thermal decomposition (NaN_3, for example) also exhibit photodecomposition. Review the suggested mechanisms for photo- and thermal decompositions of these compounds and give reasons for the simultaneous photo and heat sensitivity of these compounds. [Refs.: *J. Phys. Chem.*, **70,** 3538 (1966); *J. Phot. Sci.*, **5,** 49 (1957); *Proc. Roy. Soc. London*, **280,** 566 (1964); F. P. Bowden and A. D. Yoffe, in *Fast Reactions in Solids*, Butterworth & Co. London, (Publishers) Ltd., 1958.]

4

Electrical Properties

of Surfaces

4.1. Introduction

THE CONCENTRATION OF MOBILE CHARGE CARRIERS (electrons and diffusing ions) can, to a large extent, determine many of the physical–chemical properties of solid surfaces. The concentration of these free charge carriers, however, varies widely for materials of different types. Metal surfaces have large free electron concentrations; almost every atom contributes one electron to the lattice as a whole. For semiconductors, on the other hand, or for insulators, often less than 1 out of 10^6 atoms may contribute a free electron. Since the conductivity in an applied external electric field depends on the availability of free carriers, the surface conductivities of these different materials may vary by orders of magnitude. In addition, the temperature dependence of the carrier concentration and the conductivity may be different for different materials, depending on the mechanism of excitation by which the mobile charge carriers are created.

Under incident radiation or bombardment by an electron beam the surface emits photons or electrons or, frequently, both. The emission properties of solid surfaces differ widely, just as do their mechanisms of relaxation after excitation by high-energy radiation.

The underlying reason for the differences of the conductivity mechanisms and emission properties in the surfaces of the different materials lies in the differences in their electronic band structure. Before we discuss these properties in relation to solid surfaces it is important to outline how the electronic band structure of the solid forms and what the properties of the band structure are that give rise to the differences in the electronic and emission properties of solids and solid surfaces. For illustration we will compare the energy of a free electron, the energy levels of an electron in a one-dimensional potential well, and finally the energy levels of an electron in two symmetrical potential wells separated by a barrier. We will show that in the presence of a periodic potential barrier the energy levels split. For a solid in which atoms are arranged in a periodic manner throughout the lattice and are in close proximity (a few angstroms) to each other, the periodic electronic potential gives rise to electronic bands.

Just as the harmonic oscillator model was utilized in Chapter 3 to deduce many properties of the surface (surface heat capacity, mean square displacement etc.), the concept of electron bands and the properties of bulk solids that can be derived from the electron band theory will be utilized to deduce many of the electrical properties of surfaces. Most frequently, the properties of surfaces are not independent from the properties of the condensed phase. Since the band structure model of solids have been successful in explaining many of the solid state properties one may apply it with confidence in studies of solid surfaces.

After a brief review of solid-state properties (Section 4.3) we go on to discuss the electrical properties of surfaces of different types, the surface space charge and the various emission and recombination processes that take place at surfaces.

4.2. Electron in Free Space, in a One-dimensional Box, and in Two Potential Wells Connected by a Penetrable Barrier

The energy of a free electron E can be calculated by solving the one-dimensional time-independent Schrödinger equation, which can be written as[1,2]

$$-\frac{\hbar^2}{2m}\frac{d^2\psi(x)}{dx^2} = E\psi(x) \tag{4.1}$$

$\psi(x)$ is the wave function in one dimension, $|\psi(x)|^2\, dx$ is the probability of finding an electron between x and $x + dx$. The general solution of this differential equation is of the form

$$\psi(x) = A \exp(ikx) + B \exp(-ikx) \tag{4.2}$$

where the first term on the right side is a traveling wave in the direction x, the second term represents a traveling wave in the direction $-x$, and k is the wave vector, whose magnitude is given by

$$|k| = \left(\frac{2m}{\hbar^2}\, E\right)^{1/2} \tag{4.3}$$

and is related to the deBroglie wavelength λ by the familiar equation (Chapter 1)

$$|k| = \frac{2\pi}{\lambda} \tag{4.4}$$

A and B are constants that are fixed by the boundary conditions, i.e., the flux densities to right and left, and can be determined from the usual conditions of normalization. Since the momentum $p = mv = \hbar k$, where v is the velocity of the electron, we have, by substitution into Eq. (4.3),

$$E = \frac{\hbar^2}{2m}\, k^2 = \tfrac{1}{2}mv^2 \tag{4.5}$$

which is the classical kinetic energy of an electron. Thus the energy levels of a free electron are not quantized, but all values from zero to infinity are allowed. The situation is different for an electron enclosed in a one-dimensional box with impenetrable walls. Although the general solution to the wave equation is the same (assuming that the potential energy of the electron is zero), we have to take into account the boundary conditions. If L is the length of the box, the wave function must vanish at and beyond the boundaries:

$$\psi(L) = \psi(0) = 0 \tag{4.6}$$

Substituting Eq. (4.6) into (4.2), we see that this boundary condition restricts k to the values

$$k = \pm n\, \frac{\pi}{L} \qquad n = 1, 2, \ldots \tag{4.7}$$

The energy of an electron in a one-dimensional box is given, using Eqs. (4.5) and (4.7), as

$$E = \frac{n^2 \hbar^2 \pi^2}{2mL^2} \tag{4.8}$$

If L is on the order of atomic dimensions, the energy levels, which are now discrete and quantized, are separated by broad forbidden regions as shown in Fig. 4.1.

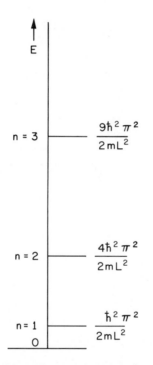

Figure 4.1. *Energy states of an electron in a one-dimensional box.*

Let us consider the effect of the proximity of another potential well (another box) on the energy levels of the electron in the one-dimensional box. We join two potential wells together in such a way that they are separated by only a small potential energy barrier. The electron can penetrate this barrier and can occupy energy states in either well. The geometry of the joint boxes is depicted in Fig. 4.2. This is similar to the case in which two atoms are brought together close enough that there is

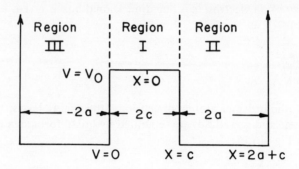

Figure 4.2. *Scheme of two potential wells joined by a penetrable potential-energy barrier.*

an overlap of their electronic wave functions. The wave equation can be written as

$$-\frac{\hbar^2}{2m}\frac{d^2\psi(x)}{dx^2} + \mathbf{V}(x)\psi = E\psi \tag{4.9}$$

where $E = \hbar^2 k^2/2m$ as given by Eq. (4.5) and the potential energy at $\mathbf{V} = \mathbf{V_0}$ may be written as

$$\mathbf{V_0} = \frac{\hbar^2\alpha^2}{2m} \tag{4.10}$$

In regions II and III the wave equation is simplified to

$$-\frac{\hbar^2}{2m}\frac{d^2\psi(x)}{dx^2} = E\psi \tag{4.11}$$

which has a general solution of the form

$$\psi_{\text{II}} = A \exp (ikx) + B \exp (-ikx) \tag{4.12}$$

In region I the wave equation is

$$-\frac{\hbar^2}{2m}\frac{d^2\psi(x)}{dx^2} + \mathbf{V_0}\psi = E\psi \tag{4.13}$$

where $\mathbf{V_0}$ is given by Eq. (4.10). Equation (4.13) has the solution

$$\psi = C \exp (i\sqrt{k^2 - \alpha^2}\, x) + D \exp (-i\sqrt{k^2 - \alpha^2}\, x) \tag{4.14}$$

Symmetry requirements fix the wave function in regions II and III and give $D = \pm C$. The wave functions and their first derivative must be continuous. We have, at $x = c$,

$A \exp(ikc) + B \exp(-ikc)$

$$= C[\exp(i\sqrt{k^2 - \alpha^2}\, c) \pm \exp(-i\sqrt{k^2 - \alpha^2}\, c)] \quad (4.15)$$

$ik[A \exp(ikc) - B \exp(-ikc)]$

$$= i\sqrt{k^2 - \alpha^2}\, C[\exp(i\sqrt{k^2 - \alpha^2}\, c) \mp \exp(-i\sqrt{k^2 - \alpha^2}\, c)] \quad (4.16)$$

At $x = 2a + c$ the wave function vanishes so that

$$A \exp[ik(2a + c)] + B \exp[(-ik(2a + c)] = 0 \quad (4.17)$$

We have three equations and three unknowns; the determinant of the coefficients of Eqs. (4.15), (4.16), and (4.17) must vanish. The solution is

$$\tan 2ka(\tan\sqrt{k^2 - \alpha^2}\, c)^{-1} = \pm \frac{k}{\sqrt{k^2 - \alpha^2}} \quad (4.18)$$

Let α approach infinity (i.e., increase the barrier height) so that $\alpha \gg k$. This modifies Eq. (4.18):

$$(\tan 2ka)(\tanh \alpha c)^{\pm 1} = \pm \frac{k}{\alpha} \quad (4.19)$$

Equation (4.19) can be rewritten in the form

$$\tan 2ka[1 \mp 2 \exp(-2\alpha c)]^{\pm 1} = \pm \frac{k}{\alpha} \quad (4.20)$$

since $\tanh x \approx 1 - 2 \exp(-2x)$ for $x \gg 1$ in the region where $\alpha \gg k$, $\tan 2ka = \tan(2ka - m\pi) \approx 2ka - m\pi$ and Eq. (4.20) yields for k, after rearrangement,

$$k = \frac{m\pi}{2a}\left\{1 \pm \frac{1}{2\alpha a}[1 + 2\exp(-2\alpha c)]\right\} \quad (4.21)$$

Equation (4.21) may be compared with Eq. (4.7). The first term on the right gives the k values for the electron in a one-dimensional box of width

$2a$ [instead of L as in Eq. (4.7)]. The k values for the electron in the two joined potential wells are shifted and give two energy states at $\pm(m\pi/4a^2\alpha)[1 + 2 \exp(-2\alpha c)]$. With respect to the energy levels that are given by Eq. (4.5), we can see that when the small probability of penetration is taken into account, any given energy level E_0 of the electron in the one-dimensional box gives two levels, E_1 and E_2.

4.3. Electrons in a Periodic Potential: The Electron Bands

A solid may be viewed as consisting of atoms, each of which is placed in potential wells in a periodic manner.[3] This is shown in Fig. 4.3. The

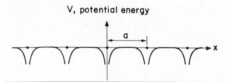

Figure 4.3. *Scheme of a solid whose atoms are placed in periodic potential wells.*

energy states of electrons of the individual atoms are all split into bands. For vanishingly small barriers between the potential wells, all bands will overlap to give a continuum of allowed levels from zero to infinity. This, of course, corresponds to the free-electron case. If the barrier between the potential wells becomes infinitely high, i.e. impenetrable, this corresponds to the case of the particle in the box. The widths of the bands decrease with increasing barrier height. In Fig. 4.4 the energy states (bands) are depicted that correspond to the free electron and an electron moving in a periodic potential by plotting the kinetic energy E as a function of the wave number k. The first four energy bands are shown, which indicate that the higher the energy of the electron, the wider is the energy band and the more similar is the E versus k curve to that for the free electron. Physically this is to be expected since the more energetic the electron, the less it is affected by the presence of a periodic potential.

One can see from Fig. 4.4 that the wave number k may vary from zero to plus or minus infinity. Since the energy states of an electron in a periodic potential are identical from period to period as shown by the periodicity of the E versus k curves, we can derive all the properties of the

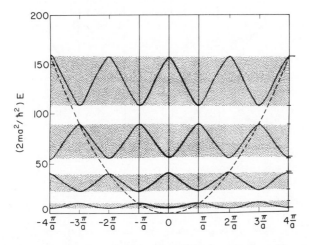

Figure 4.4. *Energy of a free electron (dotted line) and allowed energy bands (shaded areas) for electrons moving in a periodic potential as a function of wave number, k.*

electronic bands in one dimension from their properties in the first period.[3] It is customary to use the values of k in the first band as between $-\pi/a$ to $+\pi/a$. This is the *Brillouin zone*. The heavily drawn portions of the E versus k curve in Fig. 4.4 represents the band structure in this manner. Owing to the periodicity of the function E and its symmetry about $k = 0$, it follows that the derivative of E as a function of k, dE/dk, is equal to 0 at $k = 0$ and $\pm\pi/a$.

The electronic levels of the atoms in the solid are filled in accordance with the Pauli exclusion principle, which states that each quantum state can be occupied by no more than two electrons of opposite spins. Thus the maximum number of electrons that can be accommodated in any band is $2n$, twice the number of atoms. As electronic levels are filled from the lowest lying state toward the highest, we can arrive at two different conditions at the outermost electronic band: It could either be partially filled or it could be fully occupied. These conditions are shown in Fig. 4.5. These two different conditions have far-reaching implications in determining the electrical transport properties of the solid and the emission and optical properties as well.

Let us now briefly consider the motion of the electrons in the different electronic bands. In those electronic bands completely filled with electrons, the application of an external field has virtually no effect on the electron distribution. Due to the lattice potential, such an electron in a filled band is already subjected to very intense internal fields on the

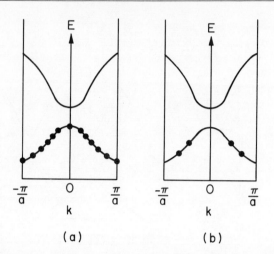

Figure 4.5. *Scheme of the outermost electron bands: (a) completely filled; (b) partially filled.*

order of 10^8 volt/cm larger than any field that could easily be applied externally across the solid. Thus an external field which is smaller than the internal field cannot significantly change the electronic band structure. Also, it is easy to see that the filled electronic bands cannot carry current. Under an applied field the k values of all the electrons may vary at the same rate; however, since all the states are occupied, the electrons just interchange positions and the field produces no net change in the occupational states. Thus the average velocity of the electrons is the same as in the absence of a field, i.e., zero. A crystal, having only bands that are completely filled or entirely empty, is a perfect insulator, but the situation is markedly different in the case of a partially filled band.[4] This is illustrated in Fig. 4.5, where the lower band is shown half filled. Since unoccupied states are now available, an external field would give rise to a net shift in the electron distribution. The new distribution would be characterized by a nonvanishing average velocity and therefore results in an electric current. The solids which are characterized by half-filled or partially occupied uppermost electronic bands show metallic conductivity and, under conditions of applied electric fields, show high conductivity at room temperatures or even at much lower temperatures than that. Thus we see how the electronic band occupancy at the topmost level can determine the conductivity of a solid and thereby allows us to classify the different solids according to their charge-carrier concentration or carrier mobility. The figures showing the partially empty and fully occupied electronic bands depict the situation at $0°K$. At this temperature the insulators show no electrical conductivity whatsoever. At higher temperatures, however, there is a finite probability of thermal excitation of the electrons across the band gap into the empty conduction band where they can move under the influence of an external electric field depending on the thermal energy $k_B T$ and the separation between the gaps of the lowest unoccupied (conduction band) and the topmost fully occupied (valence band) energy bands. If the forbidden gap is on the order of 0.05–0.5 eV, the crystal becomes highly conducting even below room temperature. On the other hand, if the band gap is on the order of several electron volts the crystal will, in effect, remain an insulator even at room temperature or at elevated temperatures. Even though we are distinguishing semiconductors from insulators, the distinction is only quantitative and is on the basis of the magnitude of their electronic conductivity. A semiconductor will become an insulator at low temperatures, whereas an insulator will become a semiconductor when the temperature is high enough. In Table 4.1 the band gaps of several semiconductors and insulators are given at room temperature.[5]

Table 4.1. *Electron Band Gaps*[5,7] *for Several Solids*

Material	E_{gap} (eV)†	Material	E_{gap} (eV)†	Material	E_{gap} (eV)†
B	1.39	GaSb	0.70	CsCl	8.5(O)
C(diamond)	5.2	InSb	0.17	LiBr	8.5(O)
SiC	2.8	ZnO	3.3(O)	NaBr	
Si	1.09	CdO	(2.2)(O)	KBr	
Ge	0.66	BaO	4.8(O)	RbBr	
α-Sn	0.08	ZnS	3.6(O)	CsBr	7.5(O)
P (yellow)	2.1(O)	CdS	2.5(O)	LiI	7.0(O)
P (red)	1.5(O)	ZnSe	2.6(O)	NaI	7.0(O)
As (grey)	1.2(O)	CdSe	1.7(O)	KI	7.0(O)
β-Sb	0.11	ZnTe	2.2(O)	RbI	7.0(O)
α-S	2.6(O)	CdTe	1.4	CsI	7.0(O)
Se (red)	1.6(O)	HgTe	0.2	AgCl	3.0(O)
Te	0.38(O)	LiF	12(O)	CuBr	2.9(O)
BN	(4.6)	NaF		AgBr	2.9(O)
AlP	3.0(O)	KF		CuI	2.8(O)
GaP	2.25	RbF		AgI	2.8(O)
InP	1.27	CsF	9(O)		
AlAs	2.3	LiCl	10(O)		
GaAs	1.43	NaCl			
InAs	0.33	KCl			
AlSb	1.52	RbCl			

† Values (in electron volts) are for 300°K, and those determined optically are designed with an (O).

4.4. Selected Electrical Properties of Solids

We shall now review many of the concepts that were developed from the application of the electronic band theory as we discuss some of the electrical properties of solids. It should be emphasized that the discussion of the solid state concepts in this section is not complete; many of the equations are given without proof and the various solid state properties are discussed only to the extent necessary to utilize them in discussions of surface electrical properties. For detailed treatments of solid state properties the interested reader is referred to several excellent textbooks that are available on the subject and are listed among the references.

4.4.1. POSITIVE HOLES, INTRINSIC AND EXTRINSIC (IMPURITY) SEMICONDUCTORS

It is worthwhile at this point to introduce the concept of positive holes. When electrons can be thermally excited into the unfilled conduction band, they leave behind vacant electronic states in the previously filled valence band.[6] This situation is illustrated in Fig. 4.6. Since in thermal equilibrium electrons tend to fill the lowest available energy levels first, the conduction electrons occupy states near the minimum of the band and the empty states in the valence band will be found near the band maximum. The electrons in the conduction band behave essentially as free electrons, in that they are characterized by a positive mass and a

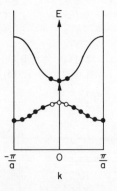

Figure 4.6. *Excitation of electrons from the filled valence band into the unfilled conduction band.*

negative charge. The electron vacancies in the valence band, however, are characterized by a positive mass and positive charge. This remarkable characteristic of a nearly full electron band with vacancies that act as positive holes is one of the most interesting features of the electronic band model. We can easily see why the electron vacancies behave as virtual positive holes. Upon the application of an external electric field we can make the electrons in the conduction band and in the valence band move along the external field. However, in the valence band as the electrons move in one direction the electron holes or electron vacancies will move in the opposite direction to give rise to a flux of seemingly positive charges. We shall see later that we can distinguish between two types of semiconductors. In the *intrinsic* semiconductors, conduction electrons may be produced by direct excitation (thermal, optical, etc.) of electrons across the band gap from the valence band to the conduction band. Under these conditions the concentration of free electrons in the conduction band equals the concentration of electron vacancies or holes in the valence band. For *extrinsic* or impurity semiconductors a large concentration of electrons in the conduction band or holes in the valence band is produced by the introduction of suitable impurities that, by ionization, provide one or the other type of mobile charge carriers in excess of their intrinsic concentration in the solid.[7] Under these conditions we can obtain semiconductors that have excess electronic conductivity, and these are often referred to as *n*-type semiconductors. Semiconductors having positive holes in their valence band in excess of the intrinsic concentration of holes due to the presence of ionized impurities are called *p*-type semiconductors.

4.4.2. THE ENERGY-LEVEL DIAGRAM

For many purposes, in analyzing the electrical properties of metals or semiconductors, we are not concerned with the detailed shape of the electronic bands. One may conveniently represent schematically the electronic bands by straight lines where the potential energy of the electron near the top of the valence band and at the bottom of the conduction band is plotted against distance x through the crystal starting from the surface ($x = 0$). The energy gap represents the minimum potential energy difference between the two bands.[6,8] In this type of diagram the electron energy increases upward and the energy of the positive hole increases downward, as indicated in Fig. 4.7. For a homogeneous crystal the bands may be horizontal, as shown in this figure.[9] We will see that at the surface the bands may vary in energy with respect to their value in the bulk of the solid since the free carrier concentrations at the surface may be different from those in the bulk of the

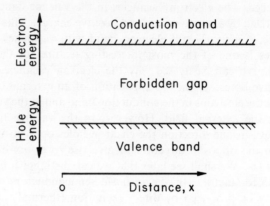

Figure 4.7. *Energy-level diagram as a function of distance x from the surface (x = 0).*

crystal. The properties of the surface space charge and of the electronic surface states that govern the bending of the electronic bands at the surface will be discussed later in this chapter.

4.4.3. ELECTRICAL PROPERTIES OF IMPURITIES IN SOLIDS

No material is ideally pure, just as no material is void of lattice defects. Most foreign elements have finite solubility in any solid, and purification techniques do not allow one to remove all the impurities completely. For a solid of average density $\rho = 5$ g/cm^3 and atomic weight $M = 100$ g/mole, the atom concentration in the usual units of atoms per cubic centimeter is $c = (\rho/M)6.02 \times 10^{23}$ atoms/mole $= 3 \times 10^{22}$ atoms/cc. If such a material is purified so that the impurity concentration is less than 1 part per million, this would correspond to an impurity concentration of about 10^{16} atoms/cc. This is near the limit of purification at present for most metallic solids. Semiconductors, however, have been purified to a much greater extent because of thorough research in their technology of crystal growth, and today one may obtain semiconductor crystals in which the electrically active impurity concentrations are at a level of 10^{12}–10^{13} atoms/cc.

The impurities in the crystal are either neutral or are ionized, releasing or capturing electrons from atoms in the host lattice. In metals the electrons provided by or captured by the impurity atoms play little role in influencing the electrical properties in general. The carrier concentration in metals is so large (since virtually every atom of the host lattice provides one electron to the conduction band which then becomes available for carrying current) that the presence of trace impurities in part-per-million concentration has no marked effect on the electrical conductivity. In semiconductors or insulators the carrier concentration is small in general at room temperature or below. Thus if impurity atoms are introduced that have low ionization energy or large electron affinity, the electron concentration could be increased or could be lowered by orders of magnitude. There are two ways in which an impurity atom can enter a perfect lattice. It can replace an atom from the host crystal, in which case it is called *substitutional impurity*, or it can occupy a position between regular lattice sites and then it is called an *interstitial impurity*. These two conditions of incorporation of impurities are shown schematically in Fig. 4.8.

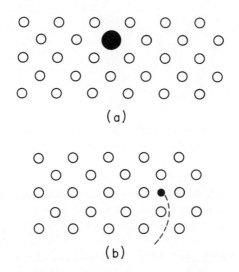

(a)

(b)

Figure 4.8. *Impurity atom in (a) substitutional and (b) interstitial lattice position.*

A description of the various mechanisms of incorporation of impurity atoms into the different crystal lattices is outside the scope of this book; however, it should be pointed out that different lattices may have many different mechanisms for incorporating impurities. In an ionic lattice the incorporation of ionic impurities of different charge types introduces excess concentrations of lattice defects such as charged cation or anion vacancies. This is necessary in most cases to maintain charge neutrality in the solid. In electronic semiconductors, on the other hand, the introduction of impurities that can donate or accept an electron from the host

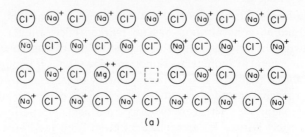

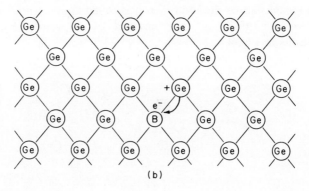

Figure 4.9. (a) *Mechanism of* Mg^{2+} *ion incorporation into a sodium chloride lattice;* (b) *mechanism of boron atom incorporation into a germanium crystal lattice.*

lattice can take place without the creation of excess charged lattice defects. Examples of these different mechanisms for incorporation of impurities in substitutional lattice positions are given in Fig. 4.9. The introduction of Mg^{2+} ion into the sodium chloride lattice gives rise to the formation of excess sodium vacancies.[7,10] The incorporation of impurities from the group III elements in the periodic table into group IV semiconductors, such as boron into germanium, can take place without the formation of such vacancies. In this case boron has only three valence electrons, while germanium has four localized electrons which participate in forming the germanium–germanium bonds. When gallium or boron substitute for a germanium atom, in order to form covalent bonds with its neighbors, it tends to accept an electron from one of the neighboring germanium atoms. This can easily be carried out at room temperature, and this process leaves behind an electron vacancy (a hole) in the germanium lattice that can move freely throughout the lattice. Upon introduction of an impurity from among the group V elements in the periodic table, such as arsenic or phosphorus, one substitutes an atom with five electrons for a germanium atom, which has only four electrons.

The fifth electron has a very low ionization energy in the germanium lattice ($\sim$0.01 eV). If the arsenic atom is ionized, this electron will move about in the crystal and become available for conduction. In terms of the electronic band model, boron in germanium introduces electron acceptor energy states in the forbidden gap (in the band gap between the valence and conduction bands). By accepting electrons from the valence band, it creates freely moving holes (electron vacancies) in that band. Alternatively, arsenic atoms donate electrons to the germanium lattice, and these can move in the germanium conduction band unimpeded, thereby contributing to the electronic conductivity.

Let us now estimate the binding energy of an electron on an arsenic ion.[5] Such an electron is moving under the influence of a Coulomb potential **V**, owing to the positive charge localized on the ionized donor,

$$\mathbf{V} = -\frac{e^2}{\epsilon r} \tag{4.22}$$

where ϵ is the effective dielectric constant of the medium, r the radius of the electron orbit, and e the charge. This electrostatic attraction is similar to that between a proton and the electron in the hydrogen atom, and thus the ionization energy to remove the electron from a hydrogen atom could be used to estimate the binding energy of an electron to the arsenic ion.

The ionization energy E of the hydrogen atom is given by $E = R_\infty n^2$, where R_∞ is the Rydberg constant and n is the principal quantum number. For $n = 1$, $E = 13.6$ eV, which is the experimentally observed ionization energy. However, there are a number of differences between the ionization of a hydrogen atom and arsenic atom in a germanium crystal. First, the attractive force acting on the electron is reduced by the dielectric constant of the host crystal by a factor of $1/\epsilon$. This, in turn, increases the radius of the electron orbit by a factor that is proportional to ϵ ($r \propto \epsilon$) so that the binding energy is reduced by a factor of $1/\epsilon^2$. Second, the "effective mass" of an electron m^* in the crystal lattice† is different from the free electron mass m. Thus the ionization energy, E, of an electron about arsenic ion in a crystal lattice can be approximated by the equation[11]

$$E \text{ (eV)} \approx \frac{13.6}{\epsilon^2} \frac{m^*}{m} \tag{4.23}$$

† The idea of "effective mass" of an electron or of an electron hole moving in a periodic potential, i.e., different from the mass of a free electron, is important and may be difficult to understand. Detailed discussion of this concept can be found in refs. 3, 4, 7, 9, and 13.

In germanium, where $\epsilon = 15.8$ and the electronic mass ratio, $m^*/m = 0.17$, the binding energy of the electron on the arsenic atom can be calculated to be 0.0085 eV. The experimental binding energy is 0.0127 eV. This model predicts that the binding energy of an electron on an impurity atom embedded in a crystal lattice is not sensitive to the properties of the impurity except its charge. Thus impurity elements from the same group in the periodic table will have similar ionization energies in a given host lattice. This model has been vindicated by experiments that determined the ionization energies of different impurities. A list of experimental binding energies is given in Table 4.2 for the silicon

Table 4.2. *Ionization Energies of Several Impurities in Silicon* ($\epsilon = 12$)

Impurities	Experimental ionization energy (eV)
Li ⎤	0.033
P	0.044
As ⎬ Donors group V	0.049
Sb	0.039
Bi ⎦	0.069
B ⎤	0.045
Al	0.057
Ga ⎬ Acceptors group III	0.065
In ⎦	0.16

host lattice. We may also estimate the radius r of the orbit of the ionized donor electron, assuming again only Coulombic interaction. This is given by

$$r = a_0 \frac{\epsilon}{m^*/m} \tag{4.24}$$

where a_0 is the *Bohr radius* of the hydrogen atom ($a_0 = 0.529$ Å) and r is then the Bohr radius in the medium. For the electron from arsenic, $r = 49$ Å. We can see that the electron can overlap many germanium atoms in the neighborhood of the arsenic ion.

As long as the electron concentration due to direct excitation from the valence band to the conduction band is small, the introduction of impurities can change the carrier concentration markedly. Incorporation of impurities is a common way of custom tailoring the carrier concentration and the type of charge carriers that exist in semiconductors and insulators. These doped semiconductors play an all-important role in semiconductor technology and in the production of a great variety of solid-state electronic devices.

4.4.4. ELECTRICAL CONDUCTIVITY

It is perhaps simplest to describe electrical conductivity as a special case of transport phenomena. Mass transport, the flux of atoms along a concentration gradient, has already been considered, when the concept of atomic diffusion was discussed. Similarly, one can write the current flux of free charge carriers, j, in the bulk solid in an electric field $E = -\nabla V$, as

$$j = -\sigma \nabla V = \sigma E \tag{4.25}$$

where σ is the constant of proportionality, the conductivity, and V is the electrostatic potential.[4,9] It is generally assumed that σ is independent of E (just as we have assumed that the diffusion constant is independent of the atom concentration). The current flux j can be written as

$$j = n_e e v_d \tag{4.26}$$

where n_e is the free carrier concentration in units of cm^{-3}, e is the unit charge ($e = 1.6 \times 10^{19}$ C), and v_d is the *drift velocity* or *average velocity* of the electrons in the solid before being stopped by a collision. If we define the *mobility* as the velocity of the charged particles in a unit electric field,

$$\mu = \frac{v_d}{E} \tag{4.27}$$

we have, by substituting Eqs. (4.26) and (4.27) into (4.25),

$$\sigma = n_e e \mu \tag{4.28}$$

It should be useful to derive the relation between the mobility and the diffusion constant of charge carriers, which are both important parameters in describing charge transport. Charge carriers, in addition to moving in an external electric field, also tend to diffuse in a concentration gradient from regions of high to regions of low charge concentration. Therefore, when current flows due to an external electric field, a diffusion current j_{diff}, which is proportional to the concentration gradient is also established in the opposite direction.[2,8] In equilibrium the two currents would be equal: $j = j_{\text{diff}}$. Combining Eqs. (4.26) and (4.27), we have $j = n_e e \mu E$. The diffusion current j_{diff} is equal to $-eD(dn_e/dx)$ if we assume that the concentration gradient exist only in the x direction. Thus in equilibrium

$$-n_e e \mu \frac{d\mathbf{V}}{dx} = -eD \frac{dn_e}{dx} \qquad (4.29)$$

Integration of Eq. (4.18) yields

$$n_e = \text{constant } e^{\mu \mathbf{V}/D} \qquad (4.30)$$

A free electron in the solid has a potential energy $-e\mathbf{V}$. If we assume that the free electron distribution in the solid in equilibrium obeys the Maxwell–Boltzmann distribution law, the free electron concentration can be expressed as

$$n_e = A e^{e\mathbf{V}/k_B T} \qquad (4.31)$$

where A is a normalizing constant. Combining Eqs. (4.30) and (4.31), we obtain

$$D = \frac{k_B T}{e} \mu \qquad (4.32)$$

This is the well-known Einstein relation[12] between the diffusion constant and the mobility of charge carriers; it is very useful in studies of ionic solids in which both the charge carriers and the diffusing species are ions. Determination of the mobility from ionic conductivity measurements yields the diffusion constant as well, by application of Eq. (4.32).

The electron can undergo scattering of many types as it moves through the lattice under the influence of the external electric field. The various scattering processes in a crystal can often be characterized by a relaxation time τ, a quantity of great importance in transport phenomena. The relaxation time represents the average time required for an electron to lose all its velocity—the implicit assumption being that at every collision the electron stops and has to start over. The mean free path l, defined as the average distance traveled by an electron before it loses all its velocity, is then given by

$$l = v_d \tau \qquad (4.33)$$

The orders of magnitude of τ and l, for metals or in semiconductors at room temperature are 10^{-12} sec and 10^{-5} cm, respectively. Using the force equation in an electric field where eE is the external force and $m(d^2V/dt^2)$ is the product of the mass and the acceleration, it can be shown[9] that the drift velocity is expressed by

$$v_d = \frac{eE}{m}\tau \qquad (4.34)$$

Substitution of Eq. (4.34) into (4.27) and (4.28) gives the mobility

$$\mu = \frac{e}{m}\tau \qquad (4.35)$$

and the conductivity

$$\sigma = \frac{n_e e^2 \tau}{m} \qquad (4.36)$$

respectively, in terms of the relaxation time. Equation (4.36) is an expression of *Ohm's law*.

For metals the carrier concentration n_e is not much different for different materials. The variations in the metallic conductivity are due to the relaxation time, i.e., the nature of scattering of the electrons in the solid. Thus for metals, μ, the mobility, is the important variable in determining the electrical conductivity.

For semiconductors and insulators there are two types of mobile charge carriers: electrons and holes that move in the conduction and in the valence bands, respectively. Therefore, the conductivity is given by

$$\sigma = (n_e e \mu_e)_{\text{electrons in conduction band}} + (n_h e \mu_h)_{\text{holes in valence band}}$$
$$= |e|(n_e \mu_e + n_h \mu_h) \qquad (4.37)$$

By convention, the current flows in the same direction as the field, E, for both electrons and holes; therefore, the electron and hole currents or conductivities are additive. Although the electron and hole mobilities, μ_e and μ_h, change as a function of the scattering mechanism, the electron and hole concentrations, n_e and n_h, are the more important parameters in determining the conductivity. By suitable doping with ionized impurities or by small changes in temperature (as we shall see below) the carrier concentrations in semiconductors can be changed by orders of magnitude.

4.4.5. TEMPERATURE DEPENDENCE OF THE ELECTRICAL CONDUCTIVITY

The free electrons in a metal are scattered because of the thermal motion of atoms and by impurities and lattice imperfections. At high temperatures, $T > \Theta_D$, where the high-temperature limit to the lattice heat capacity is approached, the mean-square displacement of atoms $\langle u^2 \rangle$, is proportional to the collision cross section.[3] From the kinetic theory the mean free path l is inversely proportional to the concentration of gas atoms and the collision cross section. Thus the electron mean free path is proportional to $l \approx 1/n_e\langle u^2 \rangle$. Remembering Eq. (3.61) for the mean-square displacement of the classical harmonic oscillator we have $\langle u^2 \rangle \approx \hbar^2 T / M k_B \Theta_D^2$. Substitution into Eqs. (4.33) and (4.36) yields

$$\sigma = \frac{n_e e^2}{m}\tau \approx \frac{e^2}{mv_d}\frac{Mk_B\Theta_D^2}{\hbar^2 T} \tag{4.38}$$

Thus at high temperatures the metal conductivity is inversely proportional to the temperature. At low temperatures $(T < \Theta_D)$, where $C_v \propto T^3$, we find that the conductivity is inversely proportional to the fifth power of the temperature: $\sigma \propto 1/T^5$. The conductivity in both high and low temperature ranges is given by the Bloch–Grüneisen equation[13]

$$\frac{1}{\sigma} = A'\left(\frac{T}{\Theta_D}\right)^5\int_0^{\Theta_D/T}\frac{x^5}{(e^x-1)(1-e^{-x})}\,dx \tag{4.39}$$

where $x = \Theta_D/T$ and A is a constant. At high temperatures, the integrand is expanded to give $\sim(\Theta_D/T)^4$, hence the temperature dependence $\sigma \propto T^{-1}$. At low temperature $(\sim 0.1\Theta_D)$ the integral has a constant value, 124.4, and the temperature dependence of the conductivity follows the fifth-power law. This strong temperature dependence is observed only for pure metals, for which impurity scattering may be neglected. There is a residual resistivity at low temperatures that is independent of the temperature and is due primarily to impurity scattering centers. Since the lattice vibrations have small amplitudes, the electron mean free path is primarily determined by the effective distance between the impurity atoms. Often, low-temperature conductivity measurements or measurement of the "resistivity ratio" for a given material (ratio of resistivities at 298°K and at 4°K in a liquid-helium thermostat) gives a figure of merit for the purity of the sample.

Thus we find that for metals at very low temperatures $(T \ll \Theta_D)$ the conductivity is virtually temperature independent because of scattering by impurity centers and lattice imperfections (point defects, dislocations, and grain boundaries). As the temperature is increased, scattering by the vibrating atoms predominates, and this type of scattering mechanism gives rise to a strong ($\sigma \propto 1/T^5$) temperature dependence at low temperatures ($T \approx 0.1\Theta_D$) and a weaker temperature dependence, ($\sigma \propto 1/T$) at high temperatures ($T \approx 0.5\Theta_D$).

For intrinsic semiconductors the free carrier concentrations (electrons and holes) depend on how many electrons can surmount the band-gap energy and jump into the conduction band. Thus the carrier concentration is an exponential function of temperature. Every time an electron jumps from the valence band into the conduction band, *two* free charge carriers (a free electron and a free hole) are created. The product of the electron and hole concentrations, n_e and n_h, respectively, is given by[7]

$$n_e \cdot n_h = N_c N_v e^{-(E_c - E_v)/k_B T} \tag{4.40}$$

where $E_c - E_v = E_{\text{gap}}$ (the band gap energy) and E_c and E_v are the relative energies of the electrons at the bottom of the conduction band and at the top of the valence band, respectively. N_c and N_v are the density of states of electrons and holes in the conduction band and in the valence band, respectively, and are given by the equations[7,8]

$$N_c = 2 \left(\frac{2\pi m_e k_B T}{h^2} \right)^{3/2} \tag{4.41}$$

and

$$N_v = 2 \left(\frac{2\pi m_h k_B T}{h^2} \right)^{3/2} \tag{4.42}$$

where m_e and m_n are the effective masses of electrons and holes in the solid, respectively. For an intrinsic semiconductor, $n_e = n_h$ and the carrier concentration is given by an equation of the form

$$n_e = n_h = \sqrt{n_e n_h} = 2 \left(\frac{2\pi k_B T}{h^2} \right)^{3/2} (m_e m_h)^{3/4} \exp\left(-\frac{E_{\text{gap}}}{2k_B T} \right) \tag{4.43}$$

The electron and hole mobilities of a semiconductor or insulator have much weaker temperature dependence than the carrier concentration. For scattering by lattice vibration, $\mu \propto T^{-3/2}$, while for scattering by impurities $\mu \propto T^{3/2}$. Just as in the case of metals for a semiconductor or insulator the scattering of electrons by thermal motion of atoms predominates at high temperatures while impurity scattering is dominant at low temperatures. However, the conductivity is dominated by the exponential temperature dependence of the carrier concentration:

$$\sigma = 2\,|e|\left(\frac{2\pi k_B T}{h^2}\right)^{3/2}(m_e m_h)^{3/4}(\mu_e + \mu_h)\exp\left(-\frac{E_{\text{gap}}}{2k_B T}\right) \tag{4.44}$$

The band gap (at 25°C) of several monatomic and diatomic semiconductors and insulators are given in Table 4.1. By using Eq. (4.44) and Table 4.1, the temperature variation of the conductivity of these materials can readily be calculated. It is customary to measure the energies E_c and E_v relative to a reference energy state that is in the band gap between the conduction band and the valence band (Fig. 4.10). This is called the

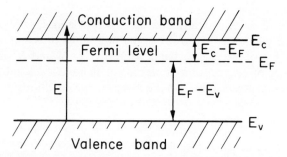

Figure 4.10. *Energy-level diagram showing the Fermi level.*

Fermi energy, E_F. By using this reference state, the free electron and free hole concentrations n_e and n_h can be expressed as[9]

$$n_e = N_c e^{-(E_c - E_F)/k_B T} \tag{4.45}$$

and

$$n_h = N_v e^{-(E_F - E_v)/k_B T} = N_v e^{(E_v - E_F)/k_B T} \tag{4.46}$$

For doped semiconductors the conductivity may no longer be controlled by the thermal excitation of electrons from the host atoms across the band gap. If the ionization energy of the impurity is smaller than that of the band-gap energy, the free electron concentration in the conduction band will be dominantly due to the ionization of the impurity centers. Let us consider a semiconductor doped with a donor impurity of small ionization energy. If the ionization energy is $E_c - E_i$, where

E_i is the energy associated with the electrons of the impurity atoms, the free electron concentration in the conduction band is

$$n_e \approx N_{imp} \exp \left(-\frac{E_c - E_i}{k_B T} \right) \tag{4.47}$$

where N_{imp} is the total concentration of impurity centers.[5] This situation is shown schematically in Fig. 4.11. Although ionized donors are left

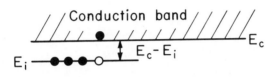

Figure 4.11. *Energy-level diagram showing electron-donor impurity states.*

behind, in general these are not mobile and do not contribute to the conductivity. Thus excitation from this kind of impurity does not yield two free carriers as does the thermal excitation of intrinsic semiconductors. Figure 4.12 shows schematically the change of conductivity as a function of temperature in an *n*-type semiconductor with a shallow donor impurity. At low temperatures the conductivity increases with increasing temperature, because of increased ionization of the donor levels. The slope of the log σ versus $1/T$ curve can give the ionization energy of the impurity centers. As the temperature is further increased, all the donor centers finally become ionized and the free electron concentration and thus the conductivity become virtually independent of temperature. At even higher temperature, thermal excitation across the band gap from the valence band into the conduction band becomes probable, and from then on the conductivity is dominated by this intrinsic process. The slope of the log σ versus $1/T$ curve now gives the band gap directly ($E_{gap}/2k$).

The temperature dependence of the conductivity for doped semiconductors may vary greatly depending on the nature (donor or acceptor), the concentration, and the ionization energy of the impurity. In case of

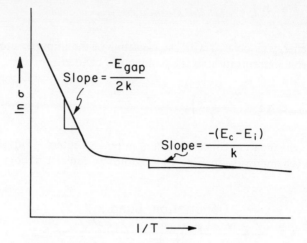

Figure 4.12. *Logarithm of electron conductivity as a function of the reciprocal temperature for an n-type semiconductor with a donor impurity.*

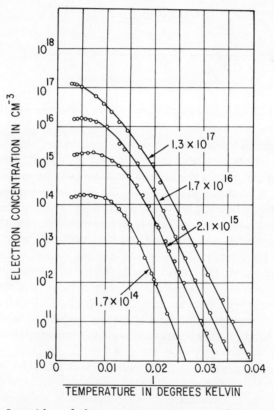

Figure 4.13. *Logarithm of electron concentration as a function of reciprocal temperature for germanium crystals doped with different atom concentrations of arsenic.*

144

more than one electrically active impurity, recombination processes between the impurity centers may make the $\sigma(T)$ curves more difficult to analyze. Detailed analysis of the effects of these various dopants on the temperature dependence of the semiconductor carrier concentration is clearly outside the scope of this book; the interested reader is referred to textbooks on the subject.[4,5,7] Finally, for illustration, we give the experimental data in Fig. 4.13 for the variation of log σ as a function of $1/T$ for germanium doped with arsenic at different concentration levels.

We would now like to discuss the electrical properties of surfaces that are the consequence of the unique anisotropic surface environment that causes a redistribution of charge density at the surface. This discussion will make use of the concepts introduced previously in the brief review of the electrical properties of solids. We will introduce and analyze the properties of the surface space charge and then the electron and ion emission phenomena will be discussed. Finally, the various excitation and recombination processes of electrons in surface atoms that provide us with information about the electronic structure and the chemical composition of atoms in the surface will be reviewed.

4.5. *The Surface Space Charge*

Mobile charge carriers at the surface (electrons or holes) can interact with adsorbed molecules and can participate in surface reactions. The flow of these free carriers toward or away from the surface can establish a potential gradient between the surface and the bulk of the material that may determine the reactivity of the surface. Any accumulation or depletion of charge carriers in the surface with respect to the bulk carrier concentration establishes a static space-charge region near the surface.[2] Such a space charge may also be induced by the application of an external electric field or by the presence of a charged layer on the surface, such as adsorbed ions or electronic surface states (to be discussed below), which act as a source or sink of electrons. The height of the surface potential barrier V_s and its distance of penetration into the bulk, d, depend on the concentration of mobile charge carriers in the surface region. In order to discuss the properties of a surface space charge, let us consider an n-type semiconductor with a bulk carrier concentration n_e(bulk).

Upon adsorption of gas molecules that are electron acceptors, mobile electrons from the solid are trapped at the surface and leave behind an "exhaustion" layer of equal concentration of immobile ionized atoms, N_D^+. In the solid *as a whole*, charge neutrality is maintained; at the surface there is a charge imbalance, however, which is shown schematically in Fig. 4.14. The region of charge imbalance is called a space-charge layer and in our model a homogeneous charge distribution is assumed throughout. In order to calculate the properties of the space-charge layer as a function of its charge density, ρ, consider a homogeneous one-dimensional solid in thermal equilibrium. The potential at any point is only a function of the distance x from the surface (where $x = 0$) and is

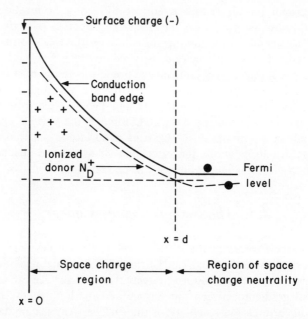

Figure 14.4. *Scheme of space-charge buildup at an n-type semiconductor surface upon adsorption of electron gas molecules.*

determined by the Poisson equation[2]:

$$\frac{d^2 \mathbf{V}}{dx^2} = - \frac{\rho(x)}{\epsilon \epsilon_0} \tag{4.48}$$

where ϵ is the dielectric constant in the solid and ϵ_0 is the permittivity of free space, a constant. In our model of the space-charge layer, $\rho(x) = eN_D^+$. Integrating twice, one obtains

$$\mathbf{V}(x) = - \frac{e}{2\epsilon \epsilon_0} N_D^+(x - d)^2 \tag{4.49}$$

At $x = d$, $\mathbf{V}(x) = 0$; i.e., d defines the distance at which the electrostatic potential due to the charge imbalance in the space-charge layer becomes zero and the electron concentration attains its bulk value again. At the surface ($x = 0$),

$$\mathbf{V}_s = \frac{e}{2\epsilon_0} N_D^+ d^2 \qquad (4.50)$$

where $\mathbf{V}_s$ is the height of the space-charge potential at the interface. Assuming that all the electrons are removed from the space-charge region and trapped at the surface, leaving behind an equal static positive charge, we have

$$e n_e(\text{bulk}) d \approx e N_D^+ d \qquad (4.51)$$

Substitution of Eq. (4.51) into (4.50) and subsequent rearrangement yields

$$d \approx \left[\frac{2\epsilon\epsilon_0 \mathbf{V}_s}{e n_e(\text{bulk})} \right]^{1/2} \qquad (4.52)$$

Thus the higher the free carrier concentration in the material, the smaller the penetration depth of the applied field into the medium. In Table 4.3

Table 4.3. *Values of d, the Distance of Space-charge Penetration for Different Values of $n_e(bulk)$ and $\mathbf{V}_s$* ($\epsilon = 16$)

$n_e(\text{bulk})$	$\mathbf{V}_s(\text{V})$	d (cm)
10^{13}	-3.5×10^{-1}	5.6×10^{-4}
10^{15}	-3.5×10^{-3}	5.6×10^{-6}
10^{17}	-3.5×10^{-5}	5.6×10^{-8}

values of d are given for various values of the bulk electron concentration and $\mathbf{V}_s$ using typical values of the constant ($\epsilon = 16$). We can see that for electron concentrations of 10^{17} cm^{-3} or larger, the space charge is restricted to distances on the order of one atomic layer or less, because the large free carrier density screens the solid from the penetration of the electrostatic field caused by the charge imbalance. For most metals almost every atom contributes one free valence electron. Since the atomic density for most solids is on the order of 10^{22} cm^{-3}, the free carrier concentration in metals is in the range 10^{20}–10^{22} cm^{-3}. Thus V_s and d are so small that they can usually be neglected. For semiconductors, or insulators on the other hand, typical free carrier concentrations at room temperature are in the range 10^{10}–10^{16} cm^{-3}. Therefore, at the surfaces of these materials, there is a space-charge barrier of appreciable height and penetration depth that could extend over thousands of atomic layers into the bulk. This is the reason for the sensitivity of semiconductor devices to ambient changes that affect the space-charge barrier height.

A space charge also exists at the surfaces of solids that exhibit ionic conductivity (NaCl, LiF, etc.) or protonic conductivity (ice, for example). These solids exhibit small carrier concentrations and are insulators in their temperature ranges of stability. During our discussion of solid surfaces with large space-charge barriers we will be emphasizing the properties of semiconductor surfaces. The arguments, however, hold for the electronic surface properties of ionic and protonic substances as well.

4.5.1. THE SPACE CHARGE AT SEMICONDUCTOR AND INSULATOR SURFACES

There is accumulation or depletion of charges at any surface in the presence of an external electric field. The space-charge layer was shown to be negligible for materials with high free carrier density ($>10^{18}$ cm^{-3}), which includes most metals. For semiconductors and insulators, however, the height of the space-charge potential barrier can be several electron volts and its penetration depth as much as 10^4 Å. There is an induced electric field at the surface under most experimental conditions because of the adsorption of gases or, as we shall see in the next part of this chapter, because of the presence of electronic surface states. Thus the electronic and many other physical–chemical properties of semiconductor and insulator surfaces depend very strongly on the properties of the space charge. For example, the conduction of free carriers across the solid or along its surface could become space charge limited. The rate of charge transfer from the solid to the adsorbed gas, which results in chemisorption or chemical reaction, can become limited by the transfer rate of electrons over the space-charge barrier. Such an adsorption process will be discussed in Chapter 5.

Let us now investigate the physical parameters that govern the height, shape, and the penetration depth of the space-charge barrier at the insulator surface. We will assume that in the absence of any space charge the electron energy levels remain unchanged right to the surface ($x = 0$) in our energy-level diagram [Fig. 4.15(a)]. However, if the surface region becomes depleted of electrons, the energy bands at the surface indicate that it would require more energy now to transfer an electron to the conduction band at the surface from, for example, the reference state E_F, on account of the space-charge potential barrier. This is shown in Fig. 4.15(b). The exponent in Eq. (4.17) is now modified to $E_c + eV_s - E_F$ at the semiconductor surface, where V_s is the height of the space-charge barrier at the surface. Conversely, it is easier now to transfer a hole to the surface since the difference between E_F and E_V becomes smaller. The

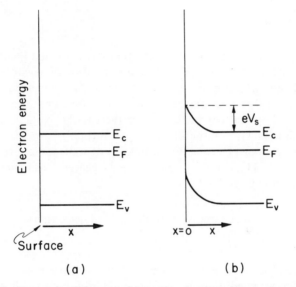

Figure 4.15. *Energy-level diagram (a) in the absence of any space charge and (b) with a surface space charge due to depletion of electrons in the surface region.*

exponent in Eq. (4.46) is changed to $E_F - (eV_s + E_v)$ at the surface. Thus we have

$$n_e(\text{surface}) = N_c e^{-(E_c+eV_s-E_F)/k_BT} = n_e(\text{bulk})e^{-eV_s/k_BT} \tag{4.53}$$

and

$$n_h(\text{surface}) = N_v e^{-(E_F-eV_s-E_v)/k_BT} = n_h(\text{bulk})e^{\,eV_s/k_BT} \tag{4.54}$$

where $n_e(\text{bulk})$ and $n_h(\text{bulk})$ are the bulk values of the free carrier concentrations (their values outside the space-charge region). [The opposite case, when there is a space-charge layer due to the accumulation of electrons at the surface, is shown in Fig. 4.19(a).] The height of the potential barrier at any point in the space-charge region, V, varies between $V = V_s$ at the surface and $V = 0$ in the bulk of the solid. The potential at any point in the space-charge layer is a function of the distance x only and is determined by the Poisson equation.[2] $\rho(x)$, the charge density at any point in the crystal, is the sum of the charges due to the mobile electrons and holes and the static positive and negative charges. The number of positive charges (static and mobile) must be equal to the number of negative charges:

$$n_h + N_D^+ = n_e + N_A^- \qquad (4.55)$$

where N_D^+ is the concentration of ions that have donated their electrons to the conduction band and N_A^- is the concentration of negative ions that have accepted (trapped) an electron. In an intrinsic semiconductor the only source of free electrons and holes is by direct excitation from the valence band to the conduction band (across the band gap). Under these conditions $N_D^+ = n_e(\text{bulk})$ and $N_A^- = n_h(\text{bulk})$, since the number of mobile charges must be equal to the number of static ionized centers that are left behind. Thus in this circumstance the charge density in the space-charge region is given by

$$\rho(x) = e \ [n_h(\text{space charge}) - n_h(\text{bulk})] - [n_e(\text{space charge}) - n_e(\text{bulk})] \qquad (4.56)$$

There are exact and also numerical solutions of the Poisson equation using different conditions of charge equilibrium. We will solve Eq. (4.48) in two limiting cases to demonstrate the overall behavior of the potential in the space-charge region.

1. The semiconductor is *intrinsic* and the imbalance between positive and negative free carriers is very small in the space-charge layer, so the height of the space-charge layer at the surface is also very small ($V \ll kT$). We can then expand the exponential functions of Eqs. (4.53) and (4.54) ($e^x \simeq 1 + x + \cdots$), and under these conditions it is permissible to neglect the nonlinear terms. For this special case the Poisson equation is given by, using Eq. (4.56),

$$\frac{d^2V}{dx^2} = \frac{e}{\epsilon\epsilon_0}\left[n_e(\text{bulk})\left(1 - \frac{eV}{k_BT}\right) + n_h(\text{bulk}) \right.$$
$$\left. - n_h(\text{bulk})\left(1 + \frac{eV}{k_BT}\right) - n_e(\text{bulk}) \right] \qquad (4.57)$$

$$\frac{d^2V}{dx^2} = \frac{e^2V}{\epsilon\epsilon_0 k_BT} \ [n_e(\text{bulk}) + n_h(\text{bulk})] \qquad (4.58)$$

Equation (4.58) can be directly integrated and yields the result

$$\mathbf{V} = \mathbf{V}_s e^{-x/l_D} \tag{4.59}$$

where l_D is given by

$$l_D = \left\{ \frac{\epsilon \epsilon_0 k_B T}{e^2 [n_e(\text{bulk}) + n_h(\text{bulk})]} \right\}^{1/2} \tag{4.60}$$

For small values of the space-charge layer, its height decreases exponentially with the penetration distance. The penetration depth, at which $x = l_D$ and the potential barrier drops to $1/e$ times its surface value $\mathbf{V}_s$, is the *effective Debye length*, which essentially characterizes the width of the space-charge region.

2. The semiconductor is *extrinsic*. Its free carrier concentration is established by the ionization of a donor (or acceptor) impurity. The concentration of ionized impurities, and hence the free carrier concentration provided by them, is much larger than the carrier concentration that can be obtained by band-to-band excitation (as in the case of intrinsic semiconductors). If only a donor (or acceptor) impurity is present, the free electron (or hole) concentration is much larger than the hole (electron) concentration. Therefore, the minority carriers can be neglected. The free electrons (or holes) can move to the surface to participate in charge transfer and become trapped, and they leave behind the static ionized donor (or acceptor) centers N_D^+ (impurity) [or N_A^- (impurity)] in the space-charge region. Under these conditions the Poisson equation can be written as

$$\frac{d^2 \mathbf{V}}{dx^2} = - \frac{e}{\epsilon \epsilon_0} [N_D^+(\text{impurity})] \tag{4.61}$$

Direct integration yields Eq. (4.49) since these experimental conditions are identical to those that were discussed in Section 4.4. Thus the height of the potential barrier $\mathbf{V}$ decreases parabolically as the square of the distance from the surfaces where the height of the space-charge barrier is $\mathbf{V}_s$. The reason for this is that the space charge consists of static charges only and this remains uncompensated by the depleted majority charge carriers (electrons here). This case corresponds to the *Schottky barrier*, which is characterized by the dependence $\mathbf{V} \approx x^2$.

It is very likely that there is accumulation or depletion of charges at semiconductor or insulator surfaces under all ambient conditions. For surfaces under atmospheric conditions, adsorbed gases or liquid layers at the interface provide trapping of charges or become the source of free carriers. For clean surfaces in ultrahigh vacuum, there are *electronic surface states* that act as traps or sources of electrons and produce a space-charge layer of appreciable height, as we will see below. Thus the mobile carriers from the surface layer are swept into the interior or are trapped at the surface as the space-charge layer is established. Therefore, the second case discussed above, where the space-charge layer consists dominantly of static charges, is the one most frequently encountered in experimental situations. For detailed derivations of the shape of the space-charge barrier as a function of penetration depth for many different conditions of charge accumulation and depletion for intrinsic and extrinsic semiconductors, the reader is referred to excellent reviews on this subject.[2,8]

We have so far considered the space-charge-layer properties only in the insulating solid, assuming that the surface layer that acts as a donor or electron trap is of monolayer thickness. This is certainly the case for many experimental systems of interest and our discussions in later chapters will be centered around them. However, considering the properties of solid–liquid interfaces or semiconductor–insulator contacts, it should be recognized that the space-charge layer may extend to effective Debye lengths on *both* sides of the interface. This is a most important consideration when one investigates the surface properties of colloid systems or of semiconductor–electrolyte interfaces.

How are the height of the space-charge barrier at the surface and other properties of the space charge measured? Changes of the conductivity of samples with large surface/volume ratio are measured as a function of various ambient conditions. Chemical surface treatments to vary the barrier height are often used; these include the exposure of the semiconductor surface to oxygen and water vapor, dry oxygen, nitrogen and water vapor, and dry nitrogen and ozone, in sequence. Water vapor on the surface gives rise to space-charge layers where electrons accumulate at the surface, while ozone and the dry gases produce surfaces depleted of electrons. The surface potential may vary as much as 0.4–0.6 volt during the treatment cycle.

4.5.2. SURFACE CONDUCTANCE MEASUREMENT

In many experiments the surface conductance $\Delta\sigma$ is measured as a function of space-charge barrier height as the space-charge layer is changed from an accumulation to a depletion layer. As the conversion from one type of space-charge layer to another type takes place, the surface conductance will have a minimum value since it is proportional to the concentration of excess charge carriers in the space-charge region[5]:

$$\Delta\sigma = e[\mu_e \, \Delta n_e(\text{surface}) + \mu_h \, \Delta n_h(\text{surface})] \tag{4.62}$$

where $\Delta n_e(\text{surface})$ and $\Delta n_h(\text{surface})$ are the excess charge carriers in the space charge:

$$\Delta n_e \,(\text{surface}) = \int_0^\infty [n_e(\text{space charge}) - n_e(\text{bulk})] \, dx \tag{4.63a}$$

$$\Delta n_h \,(\text{surface}) = \int_0^\infty [n_h(\text{space charge}) - n_h(\text{bulk})] \, dx \tag{4.63b}$$

and μ_e and μ_h are the electron and hole mobilities in the space-charge region (apparently not much different from their bulk values). Note that

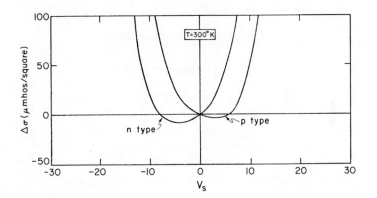

Figure 4.16. *Variation of the surface conductance as a function of the surface potential, V_s, for an n-type and a p-type semiconductor.*

the surface conductance has dimensions of ohm^{-1}, thus is independent of the surface area. The dependence of surface conductance on the space-charge barrier height can be obtained from the relationship of Δn_e(surface) and Δn_h(surface) and the surface potential V_s. In Fig. 4.16 a typical surface conductance plot is shown as a function of the height of the surface space-charge barrier for an n-type and a p-type extrinsic semiconductor. It is apparent that both electron accumulation and depletion layers are characterized by high surface conductances. The surface conductance passes through a minimum where there are very few mobile carriers in the space-charge region. The calculated dependence of the space-charge barrier height on changes of charge density in the space-charge layer for a given semiconductor (with known carrier concentration and carrier type) can now be compared with the experimental data. The discrepancy between the observed and calculated behavior can be used to learn about the distribution and density of electronic surface states (to be discussed below).

4.5.3. FIELD-EFFECT AND SURFACE CAPACITANCE MEASUREMENTS

Field-effect and surface capacitance measurements rely on the application of external electric fields to change the height of the space-charge barrier instead of using adsorbed gases for the same purpose. In a field-effect experiment the surface conductance is measured as a function of the

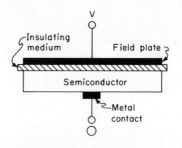

Figure 4.17. *Experimental geometry for surface-capacitance measurements.*

electric field applied to a plate parallel to the semiconductor surface and separated from it by a small gap. The circuit geometry is arranged so that the surface capacitance is much larger than the geometric capacitance. Thus the field induces charge which causes a redistribution of surface charge in the space-charge layer (i.e., the field changes the surface conductance on the order of 1 per cent of the total conductance for most experimental geometries). In surface capacitance studies the semiconductor is used as one plate of a plane-parallel capacitor separated from the other plate (a metal) by a thin insulating layer. From the change of the capacitance as a potential is applied across the capacitor, the charge density in the space-charge layer and the barrier height are deduced. This technique has been successfully used to study the electronic properties of semiconductor–liquid interfaces as well. Typical experimental geometries for this measurement are illustrated in Fig. 4.17.

4.6. Electronic Surface States

When the concept of electron bands of solids was introduced, the effect of the surface on the band structure was not considered. The appearance of allowed energy bands, which are occupied by the electrons in the solid and which are separated by forbidden energy gaps, are predicted from the solution of the Schrödinger equation for electrons that move in a periodic potential of infinite dimensions. The solid, however, is not infinite but is bounded by surfaces. In turn, surface atoms have fewer nearest neighbors and are in an asymmetric environment. The introduction of such a discontinuity as the surface perturbs the periodic potential and gives rise to solutions of the wave equation that would not have existed for the infinite crystal.[2] These are derived by using appropriate boundary conditions to terminate the crystal and are called *surface-state wave functions*. These wave functions have solutions that predict the presence of electronic energy states localized at the surface. These states can trap electrons or release them into the conduction band.

The allowed energy levels of the surface states lie in the forbidden gaps of the bulk band structure. The concentration of electronic surface states in clean surfaces can be equal to the concentration of surface atoms ($\sim 10^{15}$ cm^{-2}). Impurities or adsorbed gases have the effect of reducing the surface-state density that can be monitored by field-effect or other measurements. Two types of surface states are customarily distinguished according to the different boundary conditions used in deriving their properties. The *Tamm states* are due to the asymmetrical termination of the crystal lattice at the surface. The *Shockley states* are derived by applying boundary conditions to the wave equation that allow symmetrical termination of the lattice at the surface. These are schematically shown in Fig. 4.18.

One important consequence of the presence of electronic surface states is that the electron bands are modified at the surface even in the absence of a space charge or electron acceptor or donor species (such as adsorbed gases). The shape of the conduction band at the surface of an intrinsic semiconductor in the presence of electron-donor and electron-acceptor surface states is shown in the energy-level diagrams in Fig. 4.19(a) and (b).

Experimental investigations of the properties of electronic surface states utilize the same techniques used to study the surface space charge. Injection of charge carriers into the surface gives rise to two effects: variation of the charge density in the space-charge region and changes in the occupancy of electronic surface states. The surface conductance

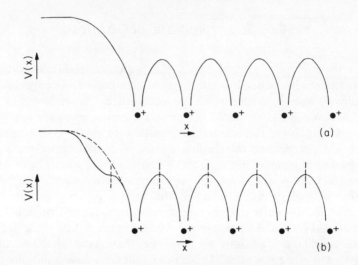

Figure 4.18. *Schematic representation of the surface boundary conditions which give rise to the (a) Tamm and (b) Shockley surface states.*

changes only owing to the appearance of mobile charges in the space-charge region; the carriers that are trapped in the surface states do not contribute to the conductivity. Thus the electronic surface states reduce the change in surface conductance upon injection of charges in a field-effect measurement. They also attenuate any change in the height of the space-charge barrier. In the presence of high-density electronic surface states the field effect may not even be detectable.

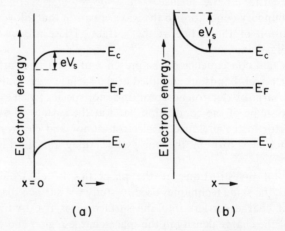

Figure 4.19. *Energy-level diagrams for an intrinsic semiconductor in the presence of (a) electron-donor or (b) electron-acceptor surface states.*

Data from surface conductance, field-effect, or capacitance measurements, when compared with predicted values for the changes in surface conductance that are computed by using a suitable model, can yield definitive information about the energy distribution and occupancy of surface states.

4.7. Emission and Recombination at Solid Surfaces

4.7.1. INTRODUCTION

If electrons in a solid are excited by high-energy incident radiation, by thermal energy, or by electron impact, several energy-transfer processes can take place. The energy may be transferred to one or more electrons which are then emitted from the surface or excited into the conduction band. This primary process is generally followed by secondary, recombination processes by which the system returns to thermal equilibrium. These recombination processes may involve emission of electromagnetic radiation or further electron emission. Since the energy of the incident beam is, in general, absorbed dominantly by surface atoms or in a thin surface layer, both the primary excitation and the recombination process may provide us with a great deal of information about the physical–chemical properties of surface atoms.

In addition, in many surface reactions the absorption of incident radiation is one of the most important, and often rate-determining, reaction steps. Examples of chemical surface reactions of this type are photodecomposition, thermal decomposition, electron beam polymerization, etc. Therefore, we will discuss the different prominent electron-transfer processes in some detail. First, we focus our attention on emission and recombination mechanisms that involve only valence electrons; these provide us with a great deal of information about surfaces and are frequently investigated by a variety of experiments. Then we will discuss emission and recombination of electrons in deeper-lying electron bands that can also be used to identify surface atoms and explore their electronic structure.

4.7.2. PENETRATION OF INCIDENT RADIATION

Let us consider the penetration depth of electromagnetic radiation that is energetic enough to cause electron excitation in the solid. The incident intensity I_0 is attenuated according to Beer's law[14]:

$$I = I_0 e^{-\alpha d} \tag{4.64}$$

where I is the intensity after penetration of a thickness d and α is the absorption coefficient. Taking a typical value for the absorption coefficient, $\alpha = 10^5$ cm^{-1}, we find that the intensity is reduced to 1 per cent of the incident intensity $[(I/I_0) = 10^{-2}]$ at a thickness $d = 4.6 \times 10^{-5}$ cm. Thus the radiation is being absorbed in a thin surface layer.

The stopping power of solids for electrons is even greater. As we have discussed in Chapter 1, low-energy electron diffraction reveals the structure of surfaces since, unlike x rays, electrons in the energy range 5–120 eV penetrate only 1–3 atomic layers at the surface. We may express the depth of penetration d of the electron beam at or near normal incidence as a function of its energy using an empirical relation,

$$d(\text{Å}) \approx 2 + \left[\frac{E(\text{eV})}{150}\right]^n \tag{4.65}$$

where the exponent varies between $n = 1$–2 for the different solids.[15] We can see from Eq. (4.65) that the penetration depth becomes greater than a few atomic layers only above 1000 eV. Therefore, in experiments with electrons of higher energy, where information about surface atoms is desired, the electron beam is aimed at the surface at a grazing angle of incidence in order to minimize penetration.

Atomic beams of thermal energy are back-reflected from the surface without any penetration. (Atomic beam scattering from surfaces will be discussed in Chapter 5.) Low-energy ions and "slow" neutrons are also used in experiments to obtain information about surfaces, but since these beams are not being used frequently in surface studies at present, they will not be discussed here.

4.8. Emission Processes Involving Valence Electrons

4.8.1. WORK FUNCTION

When a metal filament is heated in vacuum, electrons boil off its surface and can be collected on a positively charged plate placed a short distance away. This phenomenon is often called *thermionic emission*. The electrons that require the least amount of thermal energy to overcome their binding energy in the solid and evaporate are those in the high-energy tail of their equilibrium distribution in the metal. The energy distribution of these electrons can be approximated by a Boltzmann distribution,†

$$f(E) \approx e^{-\phi/k_B T} \tag{4.66}$$

where ϕ is the work function and is given by $\phi = E - E_F$. Figure 4.20

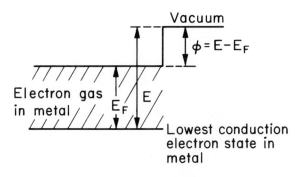

Figure 4.20. *Energy-level diagram to define the work function.*

gives the model used to define ϕ, E, and E_F. One can compute the flux of electrons of energy $E > \phi + E_F$ leaving the metal at any one temperature. This will give us the current density j (A/cm²):

$$j\left(\frac{A}{cm^2}\right) = en(E)v_z \tag{4.67}$$

where $n(E)$ is the concentration of high-energy electrons, v_z their velocity normal to the surface, and e the unit charge. After integration over the Boltzmann distribution (treating the electrons in the metal as an electron

---

* Electrons in a solid obey Fermi–Dirac statistics,

$$f(E) = \frac{1}{e^{(E-E_F)/kT} + 1}$$

where $f(E)$ gives the probability that a state of energy E will be occupied in thermal equilibrium. E_F is the chemical potential (or Fermi level) and is defined as the energy of the topmost filled electron state at absolute zero. For the high-energy tail of the distribution we have $(E - E_F) \gg k_B T$. Since under these conditions the exponential term is dominant, the unity in the denominator can be neglected and we have essentially the Boltzmann distribution,

$$f(E) \approx e^{-(E-E_F)/k_B T}$$

gas) between the limits of E and ∞ in the z direction and between $-\infty$ and ∞ in the x and y directions, we have[9]

$$j\left(\frac{A}{cm^2}\right) = \tfrac{1}{2}e\left(\frac{2k_BT}{\pi m}\right)^{1/2} N_0 e^{-\phi/k_BT} \qquad (4.68)$$

where m is the electron mass and N_0 the "density of states" that gives the number of electron states per unit volume and is given by $N_0 = 2(2\pi m k_B T/h^2)^{3/2}$. Substitution of N_0 into Eq. (4.68) gives

$$j\left(\frac{A}{cm^2}\right) = AT^2 e^{-\phi/k_BT} \qquad (4.69)$$

where $A = 4\pi e m k_B^2/h^3 = 120 \; A/cm^2 \; deg^2$. This is the well-known Richardson–Dushman equation.[16]

The electron flux leaving the surface increases with increasing temperature and decreasing work function. Thermionic emission is the most frequently used method to produce electron beams. Table 4.4 gives the

Table 4.4. *Thermionic Work Functions of Several Metals*

Metal	Temp. (°K) for 10^{-7} torr vapor pressure	Richardson constants	
		A (A/cm² deg²)	ϕ (eV)
Cs	273	160	1.81
Ba	580	60	2.11
Ni	1270	60	4.1
Pt	1650	170	5.40
Mo	1970	55	4.15
C	2030	48	4.35
Ta	2370	60	4.10
W	2520	80	4.54
Th	1800		2.7
Zr	1800		3.1

thermionic work function of several materials.[17] Barium and its compounds (oxide and silicate) and cesium are used most frequently as "cold" cathodes since large electron currents may be obtained from their surfaces even at low temperatures.

One of the advantages of using cold cathodes is that the energy spread of the emitted electrons is narrow because of the narrower Boltzmann energy distribution at low temperatures. The average thermal energy of an electron beam with a Maxwellian distribution of velocities is $\frac{3}{2}k_BT$. At the usual emitter temperatures (1000–2500°K) the largest fraction of electrons (~90 per cent) is contained in the thermal energy spread of $2\Delta E = 3k_BT$. For example, an emitter operating at 2300°K results in a beam of electrons with thermal spread ≈ 0.6 eV. Therefore, the electron beam emitted from a colder cathode is more nearly monochromatic. The higher work function of tungsten or tantalum cathodes is partly compensated by the high heating temperatures (>2000°K) that can be attained with these refractory compounds without any deterioration of their performance (they have low vapor pressures and chemically are fairly inert even at high temperatures). Refractory metals are also frequently used as electron emitters when coated with other materials of lower work function (such as thorium oxide, lanthanum hexaboride, or barium).

The work function, strictly speaking, is a bulk property—it depends on the energy distribution of the electrons in the volume of the materials. However, the measured work functions were found to be very sensitive to surface conditions. In Table 4.5 there is a list of work functions

Table 4.5. *Work Functions Measured from Different Crystal Faces of Tungsten, Molybdenum, and Tantalum*

Crystal face	Work function (eV)		
	Tungsten[23]	Molybdenum[31]	Tantalum[31]
(110)	4.68	5.00	4.80
(112)	4.69	4.55	—
(111)	4.39	4.10	4.00
(001)	4.56	4.40	4.15
(116)	4.39	—	—

measured from different crystal faces of tungsten; note that there is more than a 0.3-eV difference in the work-function values. The influence of the surface on the measured "effective work function" is due to the redistribution of the electron density at the crystal surface.[18] Because of the asymmetry of the atomic potential at the vacuum–solid interface, the energy of the electrons can be further lowered by spreading them over a larger area, as is shown schematically in Fig. 4.21. Thus the free electron at the surface has to overcome an "image potential" in addition to the work function in order to escape from the solid. The "image potential" derives from the model used to calculate this effect, which assumes the existence of an equal but opposite charge on the other side of the surface, in vacuum which is induced by the free electrons in the surface. The

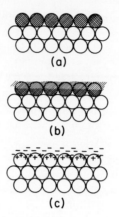

Figure 4.21. (*a*) *Electron density distribution at surface atoms that are the same as for bulk atoms;* (*b*) *redistribution of electron density at the surface;* (*c*) *the "image potential" model used to calculate the charge redistribution at the surface.*

electrostatic interaction between the electron and its positive image charge gives rise to the image potential. The image forces are, in general, increased with increasing atomic density in the surface. We find that the high-density low-index crystal faces have the highest effective work functions.

The adsorption of gases or the presence of certain impurities can markedly change the electron density distribution at crystal surfaces. Therefore, it is not surprising that the effective work function can change greatly because of the presence of foreign atoms on the surface. This effect is then used to examine the nature of chemical interaction between the adsorbed gas molecules and the solid surface. These studies will be discussed in more detail in Chapter 5.

Electron emission for measuring work function can be produced by means other than the application of supplying thermal energy by heating; for example, when electromagnetic radiation of suitable energy strikes the solid surface, electrons are emitted. A photoelectric work function may be measured by varying the wavelength of the incident light and detecting the threshold energy at which photoexcited electrons appear. Application of a large electric field (10^7–10^8 V/cm) across the surface can induce "field emission" of electrons. This technique which is also used to determine work functions will be discussed in more detail in Section 4.8.3.

Electrons are also emitted from the highest electronic energy states into vacuum during bombardment by low-energy external electrons, ions, or neutral atoms. The different methods of measuring the work function of materials are listed in several reviews on the subject.[19] We shall discuss below only one or two of the most common methods of measurements.

It is considerably easier to measure the work function of metals than the work function of semiconductors or insulators because metals have a large free-electron density, which gives a large emission current. However, the thermionic emission of insulators can also be described by a relationship similar to the Richardson–Dushman equation. The pre-exponential factor A will have a different value than that for metals since the density of state of insulators is strongly temperature dependent. The interpretation of the data is more difficult since the accumulated space charge at the surface can modify the threshold energy necessary for electron emission.

4.8.2. MEASUREMENT OF CHANGES OF WORK FUNCTION

Although the work function of surfaces can be directly determined from their electron emission characteristics, it is often not the absolute work function that we are interested in but the work-function *change* as a function of some chemical change at the surface. For example, it is commonly observed that when atoms are adsorbed on the surface, the work function changes due to redistribution of the electron density at the surface. Some of these work-function changes are small: for example, it is on the order of -0.2 eV for the adsorption of xenon on zinc, or as large as -3 eV upon the adsorption of cesium on tungsten. The stronger the charge transfer between the substrate and the adsorbed atoms or molecules, the larger is the change of work function.

The most commonly used and accurate technique of measuring the change of work function is the contact-potential-difference method. The contact potential which is the difference in work function between two clean conductor surfaces A and B, $\phi_A - \phi_B$, is changed upon adsorption of gas on one of the surfaces, A. The other reference surface, B, is kept clean during adsorption. A potential is then applied to A which is of the magnitude necessary to restore the measured contact potential difference $\phi_A - \phi_B$. This potential difference between the reference surface and the clean and covered surfaces of A gives directly the change of work function during the adsorption process. The reference electrode, B, is placed parallel to the metal or semiconductor surface to be used in this study and is mounted on a vibrating reed driven by an electromagnet[19]; this varies the plate-sample surface capacitance and gives rise to an electric signal across the detector when any potential drop (contact potential or external potential) is present between the two media.[20] The circuit includes a series of dc voltage sources adjusted to balance out the contact potential. A high-input-impedance amplifier serves as a null indicator. At balance, the voltage of the external source is equal and opposite to the contact potential difference; this way one can measure a contact potential difference as well. When gases are adsorbed or when the surface is illuminated by light of a given wavelength, the change in contact potential can be monitored within an accuracy of ± 1 mV. Another method used to measure changes in work function due to changes at the surface is the "retarding potential method." This is not as accurate as the contact-potential-difference method but is commonly used in combination with low-energy electron diffraction studies of metal surfaces. Using this method, a retarding potential is applied to the crystal that does not allow the electrons of a given energy to impinge on the surface.[21] The retarding potential is then decreased until the electron impact becomes possible. Thus this method measures the current incident on the crystal as a function of the retarding potential. When a gas is adsorbed on the surface or other effects may change the work function at the surface, the curves showing the incident electron versus the retarding potential sweep are displaced by the amount proportional to the *change* in work function. This way, the work-function change could be measured accurately because of the change in the contact potential at the surface.

4.8.3. FIELD ELECTRON EMISSION

When an electron is emitted from a metal surface as a result of thermal excitation, it has to overcome a binding energy (work function) of a few electron volts magnitude. Although this attractive potential is electrostatic in nature and therefore of long range ($\mathbf{V} \approx 1/r$), it is reduced within angstroms from the crystal surface to a fraction of its value in the solid. Thus electric fields of magnitude $E \approx 1$ V/Å or, in more common units, $E \approx 10^8$ V/cm, have to be overcome by electrons during the emission process. Therefore, another method of producing electron emission from a solid surface is by the application of a large electric field of 10^7–10^8 V/cm normal to the surface. The potential-energy barrier, which has to be overcome by an electron for it to escape by thermal excitation, can be distorted this way and allows electrons to "tunnel" through to be emitted. A potential-energy diagram for an electron at a metal surface in the absence and in the presence of such an applied field is shown in Fig. 4.22. Such an intense electric field can be obtained by applying a negative potential of 10^3–10^4 V across a small cathode tip with a radius of curvature r of 10^{-5}–10^{-4} cm, since $E \approx \mathbf{V}/r$. The current density emitted from the tip is given by the Fowler–Nordheim equation,[22]

$$j\left(\frac{A}{cm^2}\right) = 6.2 \times 10^6 \frac{(\phi/E_F)^{1/2}}{\phi + E_F} E^2 \exp\left[-6.8 \times 10^7 \frac{\phi^{3/2}}{E}\right] \qquad (4.70)$$

where E (V/cm) and ϕ (eV) are the field intensity and work function of the emitter, respectively. Equation (4.70) shows that fields of $E \approx 3$–$6 \times$

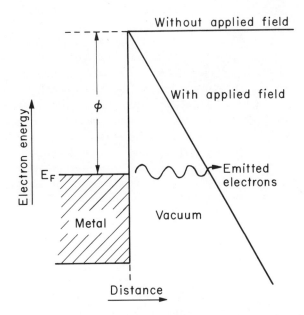

Figure 4.22. *Potential-energy diagram for an electron at the metal surface in the absence and in the presence of an applied field.*

10^7 V/cm that are used in experiments yield $j \approx 10^2$–10^3 A/cm². Using this effect, Müller has constructed a *field-emission microscope*, which allows one to study the field-emission distribution from surfaces. The electrons emitted from the cathode tip are accelerated onto a fluorescent screen (generally at ground potential), where the emission distribution of the tip is displayed with a magnification proportional to the ratio of the screen surface area to the area of the cathode tip (ratio $\sim 10^5$). Since the field-emission current is exponentially dependent on the work function between the different crystal faces, variation in the density of emitted electrons from face to face gives a characteristic image of the tip

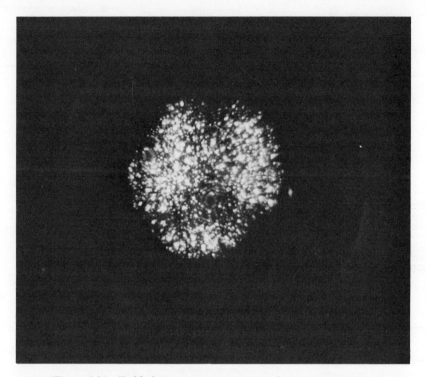

Figure 4.23. *Field-electron-emission pattern from a tungsten tip.*

surface structure with good contrast. Figure 4.23 gives a typical pattern obtained from a tungsten tip by field electron emission microscopy.

The adsorption and reaction of gases on the emitter tip can also be monitored by field electron emission since they give rise to work-function changes at the surface. This technique has also become important in studies of surface diffusion. Its application, however, is somewhat limited to those solids that can be used to fabricate cathode tips of the small radius of curvature necessary for these experiments and which are not easily destroyed by such high electric fields (because of heating effects and/or stresses caused by the field).

4.8.4. SURFACE IONIZATION: EMISSION OF POSITIVE AND NEGATIVE IONS

Consider an atom of ionization potential V_{ion} adsorbed on a metal surface of work function ϕ. If the atom is in thermal equilibrium with the solid, it may vaporize as a neutral atom from the surface after acquiring thermal energy equal to its heat of desorption from the metal, ΔH_{des}. The desorption energy necessary to vaporize it as a positive ion, ΔH_{des}^+ on the other hand, can be estimated by[23]

$$\Delta H_{des}^+ = \Delta H_{des} + V_{ion} - \phi \tag{4.71}$$

ΔH_{des}^+ is obtained by summing the energies needed to vaporize a neutral atom, ionize it in the vapor phase, then return the electron into the metal surface. If $V_{ion} - \phi$ is positive, the surface atoms are likely to desorb as neutral species since $\Delta H_{des} < \Delta H_{des}^+$. However, for systems in which the metal work function is greater than the ionization potential of the adsorbed atom, i.e., $V_{ion} - \phi < 0$, the vaporization of ionic species will occur preferentially. Thus for studies of surface ionization, high-work-function metals (W, Pt) and adsorbates with low ionization potentials (Cs, Rb, K) are utilized.

The degree of ionization, the ratio of ion flux j_+ to the flux of neutral atoms j_0 desorbing from the metal surface, is given by the Saha–Langmuir equation[24]

$$\frac{j_+}{j_0} = \frac{g_+}{g_0} \exp\left[-\frac{e(V_{ion} - \phi)}{k_B T}\right] \tag{4.72}$$

where g_+/g_0 is the ratio of the statistical weights of the ionic and atomic states. In Table 4.6 we give the calculated values for the degree of

Table 4.6. *Calculated Values for the Degree of Ionization for Different Alkali Metals on a Tungsten Surface at Different Temperatures*

T (°K)	Li V_{ion} = 5.40 eV	Na V_{ion} = 5.12 eV	K V_{ion} = 4.32 eV	Rb V_{ion} = 4.10 eV	Cs V_{ion} = 3.88 eV
1000	1.8×10^{-5}	5.0×10^{-4}	6.3	103.9	790.0
1500	5.5×10^{-4}	5.0×10^{-3}	2.2	35.8	72.0
2000	3.0×10^{-3}	1.5×10^{-2}	1.6	11.4	19.9
2500	8.4×10^{-3}	3.2×10^{-2}	1.3	7.0	9.8

ionization for different alkali metals on a tungsten surface at different temperatures. We can see that the elements with small ionization potential give dominantly ion fluxes (Cs and Rb), while for elements with large ionization potential, such as lithium and sodium fluxes, the neutral species predominate.

Equation (4.72) was verified by experiments using several different metal surfaces. Deviations from the predicted ion flux are due to the presence of impurities on the metal surface which may change its work function, and the fact that thermal equilibrium may not be completely established between the adsorbate and the surface within its residence time on the metal.[23] This latter effect can give rise to a partial reflection of the incident vapor atoms as neutral species thereby reducing the ion flux to below a value predicted by Eq. (4.72). It should be noted that the surface temperature in surface ionization experiments should be high enough so that thermal desorption of the adsorbed species can take place rapidly. Otherwise, accumulation of the adsorbate on the surface would impede the surface ionization reaction by reducing the concentration of surface sites on which ionization is to take place.

Since the metal surface is heterogeneous, there are local variations of the work function along the crystal surface. For a polycrystalline substrate the work function changes from crystal face to crystal face. Therefore, it is of advantage in surface ionization experiments to establish conditions that allow surface diffusion of the adsorbed species to occur—this way the ionization probability may be increased.

In addition to alkali metals, alkali halides (NaCl, LiF, etc.) and alkali earth metal atoms (Ba, Mg, etc.) have also been ionized by surface ionization. Tungsten and platinum surfaces are used most frequently in these studies.

The emission of negative ions has also been observed under conditions of surface ionization. If the electron affinity S of a negative ion is defined by the reaction $A + e \xrightarrow{-S} A^-$, the desorption energy of negative ions, ΔH_{des}^-, can be estimated by[23]

$$\Delta H_{\text{des}}^- = \Delta H_{\text{des}} - S + \phi \tag{4.73}$$

Here we form the negative ion by vaporizing a neutral atom and an electron from the surface and then combining them in the vapor. The electron affinities of several elements that exhibit the largest positive values are shown in Table 4.7. For these elements, negative surface ionization should be the most probable since it is a spontaneous process. For other elements, however, the electron affinity is negative; these are

Table 4.7. *Electron Affinities of Several Elements Which Exhibit the Largest Positive Values*

	F	Cl	Br	I	O	S
Electron affinity (eV)	3.6	3.7	3.5	3.2	3.1 (2.3)	2.4

not likely candidates for negative surface ionization. If $(-S + \phi) < 0$, i.e., the electron affinity is of greater magnitude than the work function (which is always positive), the atoms adsorbed on the metal surface are most likely to desorb as negative ions. The degree of ionization is given by the equation

$$\frac{j_-}{j_0} = \frac{g_-}{g_0} \exp\left(-\frac{e(-S + \phi)}{k_B T}\right) \tag{4.74}$$

which is similar to Eq. (4.54). Negative-ion emission would require metal surfaces with relatively low work function. The negative surface ionization process has been studied to a lesser extent than positive-ion emission. These studies should be somewhat more difficult to carry out since the negative-ion flux and the flux of electrons which may be emitted simultaneously from the surface thermally would have to be separated and identified.

4.8.5. FIELD IONIZATION

The first ionization energy of free atoms of most elements in the periodic table is in the range 4–25 eV. Therefore, just as for an electron in a solid, a valence electron in a free atom has to overcome an electric field on the order of $E \approx 10^8$ V/cm to be removed from the atom to infinity. If we can apply an electric field of the same order of magnitude at a metal tip, we should be able to ionize free atoms of all kinds that approach the surface. For field ionization, however, the surfaces should be charged positively in order to repel the ions formed at the surface. As the electron–ion pair (i.e., the free atom) approaches the surface, field ionization will occur at a critical distance, x_c, at the surface (about 4–8 Å), which is defined by[25]

$$x_c \approx \frac{\mathbf{V}_{\text{ion}} - \phi}{eE} \tag{4.75}$$

The model of field ionization leading to the expression in Eq. (4.75) should not hold too well for very short distances from the ionizing tip since it assumes a smooth surface, a one-dimensional tunneling process, and the classical image potential. However, it allows an order-of-magnitude estimate of the important field ion emission parameters.

The time τ necessary to ionize an atom in a given electric field E can be estimated from[26]

$$\tau(\text{sec}) = 10^{-16} \exp\left(\frac{0.68\mathbf{V}_{\text{ion}}}{E}\right)^{3/2} \tag{4.76}$$

where E is given in volts per angstrom and $\mathbf{V}_{ion}$ in electron volts. Now the probability of ionization $\mathbf{P}(t)$, during time t, is given by

$$\mathbf{P}(t) = 1 - e^{-t/\tau} \tag{4.77}$$

The probability of ionization should be expressed as a function of atomic position from the emitter surface since the atom is approaching the surface with a velocity v:

$$\mathbf{P}(t) = 1 - \exp\left[-\int_{x_a}^{x_b} \frac{dx}{v\tau(x)}\right] \tag{4.78}$$

By using Eq. (4.78) it is possible to compute the probability of ionization as a function of distance from the surface of the field tip. This was carried out by numerical integration for several gases. Figure 4.24 gives the logarithm of the ionization probability of an atom as a function of distance from the surface while traveling a distance of 1 Å toward the surface with a constant velocity. There is a series of curves for different values of the applied electric field. We can see that for fields with $E > 10^8$ V/cm, the ionization probability rapidly approaches unity.

The ion current I leaving the proximity of the field ionization tip that is maintained at a voltage sufficient to ionize all the arriving atoms or molecules ($E \approx 2 \times 10^8$ V/cm) can be established from[26]

$$I = \frac{4\pi r^2 P}{\sqrt{2\pi m k_B T}}\left(\frac{\alpha E^2 \pi}{2k_B T}\right)^{1/2} \tag{4.79}$$

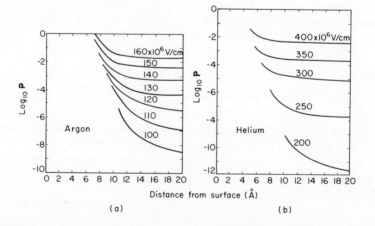

Figure 4.24. *Ionization probabilities of (a) argon and (b) helium as a function of distance from the surface at various applied electric fields.*

where P is the gas pressure, m the weight of the gas atom, α the polarizability of the atom, and r the tip radius. The current is obtained in amperes if P is in dynes per square centimeter. The logarithm of the experimental ion currents as a function of ionizing field E is given in Fig. 4.25 for different noble gases at $P = 10^{-3}$ torr and $r = 1000$ Å.

Field ionization appears to be an attractive technique to ionize molecules without appreciable fragmentation[27]; they would otherwise easily fragment during electron impact ionization in a conventional mass spectrometer. By using this technique, the ionization efficiency can be increased over that commonly encountered in a crossed electron–atomic beam ionization process or under conditions of thermal surface ionization. Although there have been exploratory studies made, this promising technique for ionizing molecules awaits future exploitation.

4.8.6. FIELD-ION MICROSCOPY

The geometry of a field-ion microscope is similar to that of a field-emission microscope. The ionizing gas most frequently used in these studies is helium, for reasons mentioned below. The helium atoms ($V_{ion} = 24.47$ eV) present at pressures between 10^{-4} and 10^{-3} torr are ionized at the field-emission tip (now at positive potential with respect to the screen which is grounded) using fields on the order of $E \approx 10^{8}$ V/cm.

The electric field strength at the surface is modulated by the atomic structure. This variation, in turn, modulates the ion current density leaving from the proximity of different crystal faces that make up the emitter. The positive ions are then repelled and accelerated onto the fluorescent screen where the greatly magnified image of the tip surface is displayed. Although the process is similar to that which gives the image of the surface in field electron emission, the use of ions instead

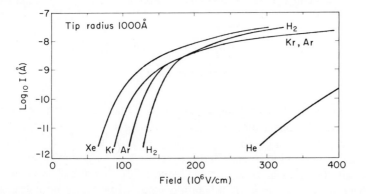

Figure 4.25. *Logarithm of ion currents as a function of ionizing electric field for xenon, krypton, argon, hydrogen, and helium.*

of electrons to form the image of the tip increases the resolution markedly. The minimum resolvable distance δ on the emitter surface is given by[26]

$$\delta = \left[\frac{2\hbar\beta r}{(2meV)^{1/2}} + \frac{16\beta^2 r^2 k_B T}{eV} \right]^{1/2} \tag{4.80}$$

where β is a small geometrical correction factor to take into account the deviation of the shape of the tip from a perfect sphere ($\beta \approx 1.5$), all other terms have their usual meaning. According to the first term, for a given tip size (which is restricted due to experimental considerations) the larger the mass of the emitted particle, the better the resolution. Since a helium atom is heavier than an electron by a factor of 1000, its use greatly improves the resolution. Owing to the larger mass, there is little displacement of ions along the surface of the tip (tangential displacement) as compared with electrons during their residence time at the emitter. Tangential displacement could introduce uncertainty in the ion position and, therefore, blurring of the image. In the second term we have the thermal energy of the ionized atom which increases the positional uncertainty of the ion because of the tangential component of its velocity. Therefore, it is of advantage to cool the emitter tip to as low a temperature as feasible and allow thermal accommodation (cooling) of the ions before emission occurs. Field-ion-microscope experiments are generally carried out at $T \approx 21°K$. At these low temperatures there are not many gases (only He, H_2, Ne, D_2) that have high enough pressures (10^{-3} torr) to produce high-intensity high-contrast images. Using He gas at $21°K$ in an ionizing field of $E \approx 4 \times 10^8$ V/cm and an emitter radius $r = 10^{-5}$ cm, the predicted resolution is $\delta \approx 1$ Å. The experimentally attainable resolution is $\delta \approx 2$–3 Å. Thus single atoms or vacancies can be identified on the different crystal faces of the tip and their movement can be monitored.

4.9. Recombination Processes and Oscillation of Valence Electrons

When an electron in an insulator or semiconductor is excited by thermal energy transfer (electron impact or electromagnetic radiation) across the band gap into the conduction band, it remains there until it loses its energy by some kind of deexcitation process.[5] The lifetime of the excited electron (relaxation time) in the conduction band can vary widely depending on the type of relaxation mechanism that dominates. For most intrinsic semiconductors it is in the range $\tau = 10^{-10}$–10^{-2} sec.

Deexcitation may take place in several ways. (1) One way is by direct band-to-band transitions. This is shown schematically in Fig. 4.26. If this recombination process dominates, electromagnetic radiation of band-gap energy is emitted. This radiative recombination is characteristic of many compounds in the II–VI and III–V groups in the periodic table. Many of these diatomic solids are used as light-emitting phosphors in cathode-ray tubes (oscilloscope screen or television screen, etc.) if the band-gap energy is in the range of visible light energy (e.g., ZnS and CdS). Phosphors that emit in the infrared are also widely utilized (GaAs, GaP, and InSb). Photon emission may also take place if (2) recombination occurs via impurity centers. In this process the electron from the conduction band is trapped by an impurity center. The energy given off as radiation corresponding to this transition is less than the band-gap energy. For example, thallium-doped sodium iodide crystals are presently used as scintillation counters. High-energy radiation causes recombination in the NaI crystal through the Tl^+ impurity centers. The radiation is then converted into an amplified electric signal by the use of an electron multiplier at the back surface of the crystal. Doping with suitable impurities can often be used to change the wavelength of the radiation from the radiative recombination process. Often more than one impurity center may partake in the recombination.

There are other indirect deexcitation processes. The electron energy can be transmitted to the crystal lattice as a whole via (3) excitation of lattice vibrations (phonon emission). The transfer of energy to the crystal requires the excitation of many phonons since the lattice vibration energies are much smaller than the electron energy to be transferred. The solid will heat up because of this process. (4) The electron that

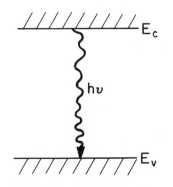

Figure 4.26. *Scheme of radiative deexcitation by electron transition from the conduction to the valence band.*

recombines may also transfer its energy to another electron in the conduction band, which may then be emitted from the solid (Auger process). Since Auger transitions are among the most probable characteristic energy-loss processes that take place during excitation by an electron beam, they can be conveniently used to analyze the surface composition and to identify surface impurities. This technique plays an important role in surface science, and we shall discuss it in more detail later in this chapter.

The dominant mechanism of energy release for a given solid depends on the relative magnitudes of the cross sections of the different recombination processes. These can vary from 10^{-22} to 10^{-12} cm². The commonly reported value for the cross section of a localized recombination center (ionized impurity for example) is about 10^{-15} cm², which is the magnitude to be expected from the physical dimensions of an atom or an ion. The recombination cross section may be calculated by assuming only electrostatic interaction between the capturing center and the free electron.[5] If we assume that the electron will be captured if it approaches the trapping center close enough so that the binding energy is equal to or greater than kT, we can write

$$\frac{e^2}{\epsilon r} \geqslant k_B T \tag{4.81}$$

Rearranging Eq. (4.83), we have for the cross section,

$$\pi r^2 \approx \left(\frac{e^2}{k_B T}\right)^2 \frac{1}{\epsilon^2} \pi \tag{4.82}$$

At room temperature (300°K) and for reasonable values of the dielectric constant ($\epsilon = 15$) a capture cross section of $\sim 10^{-13}$ cm² can be estimated from this simple model.

In a solid the nearly free electrons are known to oscillate with a characteristic frequency ω_p, called plasma frequency, which depends on the electron density. This collective oscillation of the electrons can be detected by the characteristic energy loss of an electron beam, $\Delta E = \hbar \omega_p$, which is transmitted through a thin film of the solid. At the surface due to the sudden termination of the bulk phase, one may expect a different plasma frequency for fluctuating charges. Indeed, "surface plasmons," which oscillate with a frequency ω_p (surface) that is less than the characteristic bulk plasma frequency ω_p, have been detected from the characteristic loss spectrum of the electron beam, which was studied as a function of incident angle.[28] Since the surface plasma frequency is sensitive to chemical changes at the surface (oxidation or presence of impurities) and surface roughness, its study provides a great deal of information about a variety of important surface properties.

4.10. Emission and Recombination Involving Inner Electron Shells

When a high-energy electron beam (10^3–10^5 eV) or high-energy electromagnetic radiation (x ray) is allowed to strike the solid surface in addition to electron emission from the valence band, electrons are excited from inner electron shells as well.[29] The two primary inner-shell excitation processes are illustrated in Fig. 4.27 The notation we have adopted is that most commonly used in atomic spectroscopy: the $K, L, M, \ldots$ shells refer to those with principal quantum number $n = 1, 2, 3, \ldots$ and the subscripts (L_I, L_{II}, L_{III}) indicate the multiplicity $\mathbf{j}$, which is a vector sum of the angular momentum $\mathbf{l}$ and the spin quantum number $\mathbf{s}$; $\mathbf{j} = \mathbf{l} \pm \mathbf{s}$. [For example, for sodium with a ground-state electron configuration of $1s^2 2s^2 2p^6 3s^1$ we have the $K(1s_{1/2})$, $L_I(2s_{1/2})$, $L_{II}(2p_{1/2})$, $L_{III}(2p_{3/2})$, and $M_I(3s_{1/2})$ electronic levels with experimentally distinguishable binding energies.]

An electron may be ejected from the K shell into vacuum. For sodium the minimum energy required for this process is 1072 eV. If the energy of the incident radiation is greater than the binding energy of the electron, this process, which is commonly called *photoelectron emission*, can take place. From the knowledge of the incident beam energy and by suitable analysis of the ejected photoelectrons, the electronic binding

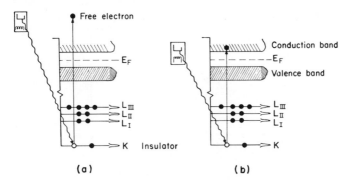

Figure 4.27. *Energy-level-diagram representation of (a) photoelectron emission and (b) x-ray absorption.*

energies in the different bands can be obtained. On the other hand, the electron may only be excited into the conduction band of an insulator or just above the Fermi level in a metal. In this case the energy of excitation is absorbed by the atoms. By variation of the energy of the incident beam the absorption spectrum characteristic of a given solid can be obtained, and from the absorption peaks at different energies the electronic binding energies can again be determined.

Energy analysis of the emitted photoelectrons is called *photoelectron spectroscopy;* the energy analysis of the absorption peaks is called x-ray absorption spectroscopy.[28] In general, the photoelectron spectra give peaks of better energy resolution (commonly ± 1 eV) than the absorption spectra. The kinetic energy of the photoelectron can be well defined (<1 eV), and any energy spread is due to the natural width of the electron band from which the electron has been ejected. In the absorption process the energy resolution of the absorption peaks depends on the width of both the inner shell and the broad unoccupied band into which the electron is excited.

Let us now turn our attention to the dominant recombination or deexcitation processes which follow the excitation of electrons from inner shells. These are illustrated in Fig. 4.28. The first mode of de-excitation is the *Auger process*, which leads to further electron emission. The second type involves the emission of electromagnetic radiation and is commonly called *x-ray fluorescence*. In the Auger transition the electron vacancy

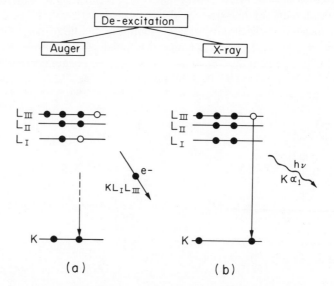

Figure 4.28. *Energy-level-diagram representation of deexcitations by (a) Auger electron emission and (b) x-ray fluorescence.*

(in the K shell, for example) is filled by an electron from an outer band (L_I shell). The energy released by this transition is transferred to another electron in one of the electron levels (L_{III} shell), which is then ejected. Energy analysis of the emitted electrons will give differences in binding energy between electronic bands participating in the Auger process (KL_IL_{III}) that are characteristic of a given element. Analysis of the x-ray fluorescence spectra gives similar information. It has been found, however, that for light elements the probability of Auger transitions (electron emission) is much greater than the probability of radiative deexcitations (x-ray fluorescence). In Fig. 4.29 the Auger electron and x-ray fluorescence yields are given for K-shell vacancies as a function of atomic number Z. It can be seen that the Auger process is dominant until $Z = 33$ (As) and then the x-ray fluorescence yield slowly increases at the expense of Auger electron emission.

In recent years Auger electron spectroscopy and photoelectron spectroscopy have come to play dominant roles in studies analyzing the composition of surfaces.[28,29] When the excitation of Auger transitions is carried out by an electron beam under a grazing angle of incidence, most of the Auger electrons are the products of electronic transitions in the topmost atomic layer. Thus this technique can conveniently be used to determine nondestructively the composition of the surface and changes of the surface composition under a variety of experimental conditions. Since the Auger transition probabilities are large, especially for elements of low atomic number, surface impurities in quantities as little as 1 per cent of a monolayer ($\sim 10^{13}$ atoms/cm^2) may be detected. In photoelectron spectroscopy, x-ray excitation is used in most studies to produce photoelectrons even though electron beam excitation could be of

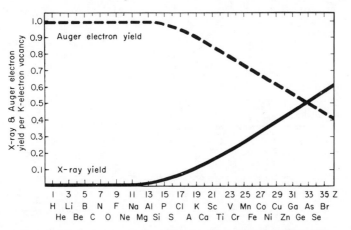

Figure 4.29. *Auger electron emission and x-ray fluorescence yields in the presence of K-shell electron vacancies as a function of atomic number.*

advantage in surface studies. Although the incident x-ray beam penetrates deeper into the solid than an electron beam of similar energy would, most of the photoelectrons ejected must be generated near the surface to be able to leave the crystal. Therefore, their energy analysis under suitable conditions can provide surface chemical analysis.

Since these techniques are important in surface science at present and promise to be even more important in the future, we will describe the techniques and the data analysis in more detail below.

4.10.1. AUGER ELECTRON SPECTROSCOPY

An experimental apparatus frequently used at present is shown schematically in Fig. 4.30. It utilizes the geometry of the low-energy electron-diffraction apparatus (the postacceleration type).[30] The fluorescent screen can be used as a collector while the grids are used for electron-energy analysis. The advantage of this system is that both Auger spectroscopy and low-energy electron diffraction studies can be carried out on the same crystal surface by using the electron optics in two different modes (Auger or LEED), alternately. An auxiliary electron gun at grazing angle incidence provides the incident exciting electron beam at energies usually 2–3 keV. Energy analysis is carried out by changing the

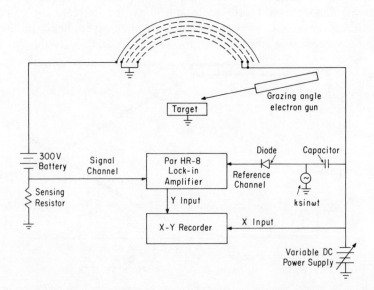

Figure 4.30. *Schematic diagram of Auger spectroscopy apparatus.*

retarding potential on the center grid (or grids) in the range of zero to incident beam energy. The current flux that appears at the collector as a function of retarding potential is shown schematically in Fig. 4.31(a). Since the retarding grid permits all electrons with energies greater than the retarding voltage to strike the collector, the high-intensity background makes the detection of an electronic transition at a well-defined energy difficult. Separation of the Auger peaks from the background of secondary electrons is carried out by superimposing a small ac signal (1- 10-V amplitude) on the retarding dc potential. Suitable differentiating circuitry and the use of a reference signal (by the use of a lock-in amplifier) allows the monitoring of the first and the second derivatives of the collector current as a function of the retarding potential: dI/dV and d^2I/dV^2. These are shown schematically in Fig. 4.31(b) and (c). In this way the Auger peaks or other energy-loss peaks become easily distinguishable from the background.

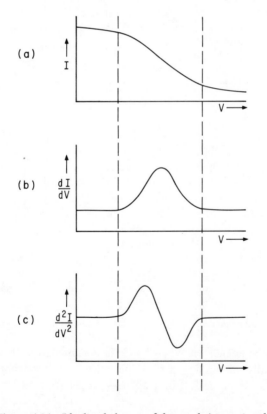

Figure 4.31. *Idealized shapes of detected Auger signals.*

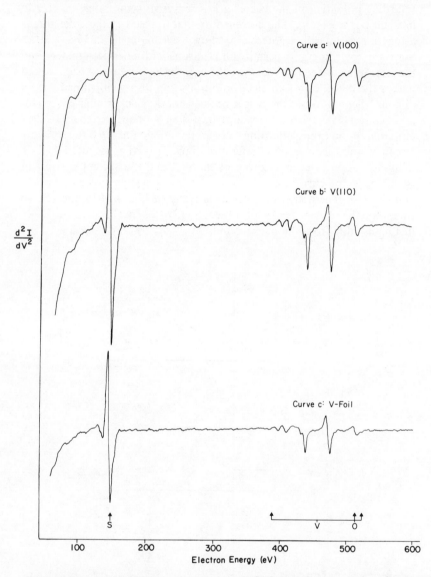

Figure 4.32. *Auger electron emission spectra from vanadium crystal surfaces.*

The energy at which an Auger peak is detected in such a spectrum, E_{obs}, is actually the binding-energy difference of the electronic shells that participate in the process. For a $KL_I L_{III}$ Auger transition we have

$$E_{\text{obs}} = E_K - E_{L_I} - E_{L_{III}} - \phi_a \qquad (4.83)$$

where ϕ_a is the work function of the analyzer grid. Since the electronic binding energies are tabulated, in most cases inspection of these tables allows one to determine the element responsible for the energy loss and the particular electronic transitions that took place. By suitable calibration with known standards, the intensities of the peaks can be used for quantitative as well as qualitative surface analysis.

It should be noted that since the Auger transition is a secondary recombination process, the peak width is independent of the energy distribution of the incident electron beam. It should be apparent from Eq. (4.83) that the characteristic Auger transition is not a function of the properties of the incident beam.

A typical Auger spectrum of a vanadium surface is shown in Fig. 4.32. The presence of small concentrations of carbon and sulfur is easily discernible. Since both Auger electron and photoelectron emission are atomic properties, these techniques can be applied to studies of solid surfaces with various degrees of crystallinity (foil, crystal, dispersed particles, etc.) and to studies of liquid surfaces as well.

4.10.2. PHOTOELECTRON SPECTROSCOPY

The apparatus commonly used is shown schematically in Fig. 4.33. A monochromatic x-ray beam is incident on the sample and generates the

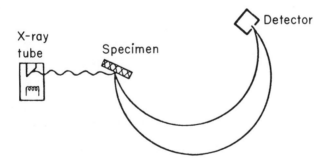

Figure 4.33. *Experimental geometry used in photoelectron spectroscopy.*

photoelectrons.[28] The emitted electrons (which include Auger electrons as well) are energy-analyzed by using a magnetic double-focusing instrument. In this way an energy resolution of $(\Delta E/E) \approx 5 \times 10^{-4}$ can be obtained at around 1 eV. The high resolution is important in these investigations since the energy of the ejected photoelectron, E_{obs}, depends on the incident beam energy $E_{x\text{-ray}}$ and is given by

$$E_{obs} = E_{x\text{-ray}} - E_b - \phi_a \tag{4.84}$$

where E_b is the electron binding energy in the shell from which it is ejected and ϕ_a is the work function of the analyzer. The electronic binding energy is thus directly calculable from Eq. (4.84). In Fig. 4.34 a typical photoelectron spectrum is shown. The intensity of the peaks can be used to determine relative or absolute concentrations of the different elements.

4.10.3. CHEMICAL SHIFT

In addition to the detection of elemental chemical composition, photoelectron spectroscopy can distinguish between the different oxidation

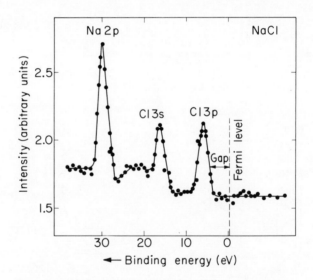

Figure 4.34. *Photoelectron spectrum of sodium chloride.*

states of elements, because of a shift in the binding energies of inner-shell electrons upon the change of valency.[28] This is commonly called the *chemical shift;* in Fig. 4.35 an example of this shift is given. It was found that the binding energy of the 1s electrons in sulfur shifts by more than 6 V as a function of the oxidation state of the sulfur. The binding energy of beryllium-1s electrons shifts to higher values by about 2.9 eV

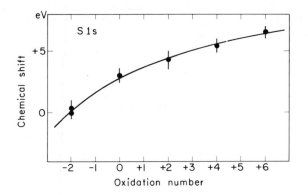

Figure 4.35. *Chemical shift of the 1s electron shell in sulfur atoms as a function of oxidation state in inorganic sulfur compounds.*

upon oxidation. This shift in the inner-shell binding energies can be rationalized if we view the atom as a charged shell of radius r. If charges are removed from the valence shell (or added to it), the electrical potential inside the shell has changed. The removal of q charges from the atom to infinity decreases the potential energy of the remaining electrons by the amount

$$\Delta E = \frac{q}{r}$$

Thus the binding energy of the core electrons should be increased by ΔE. Calculations of the chemical shifts of inner shells using more sophisticated models all indicate a shift in binding energy of 2–5 eV upon changes of oxidation states. The experimentally detectable shifts are indeed of this magnitude.

By monitoring chemical shifts, photoelectron spectroscopy has not only been able to distinguish between different oxidation states but also to detect other changes in the chemical environment that lead to the redistribution of the electron density. It could differentiate between BeO and BeF_2, for example. Carbon atoms that are surrounded by hydrogen atoms (methyl group) or are part of a carboxyl group are also distinguishable. Thus analysis of complex organic compounds with direct identification of the different constituent groups from their different chemical shifts became possible.

Auger spectroscopy can also be used to analyze the oxidation state or the changing electronic environment of surface atoms as indicated by the chemical shifts of the characteristic Auger peaks. Since the chemical shifts of the Auger transitions have been detected only recently, experimental data have not been available to such an extent as in photo-electron spectroscopy. Nonetheless, it is expected that chemical shifts will be used to distinguish between surface and bulk atoms and to monitor the changing chemical environment of surface species.

Summary

The formation of electronic bands in solids is a consequence of the periodic arrangement of atoms in close proximity to each other. The solid is an insulator or a metal if the outermost electronic band is fully occupied or partially filled, respectively. A space charge layer is created at the surface as mobile charge carriers from the solid are trapped by adsorbed molecules or electronic surface states. The height of the surface potential barrier and its distance of penetration into the bulk depend on the free carrier concentration. Emission of electrons from the valence band by thermal excitation (work function measurements) or by the application of large electric fields (field electron emission) is sensitive to the atomic surface structure and to the presence of adsorbed gases. Ionization of incident atoms at surfaces can take place under appropriate experimental conditions. Surface ionization in the presence of large electric fields (field ionization) is used for studies of the atomic structure of surfaces. Emission of electrons from inner electron shells (photoelectron emission) and electron emission during the de-excitation of surface atoms with an electron vacancy (Auger process) are utilized to determine the composition of surfaces. Energy analysis of the emitted electrons (photoelectron and Auger electron spectroscopy), in addition to determining the chemical composition, detects the oxidation state and monitors the changing chemical environment of surface atoms (chemical shift).

References

1. H. Strauss: *Introduction to Quantum Mechanics*. Prentice-Hall, Inc., Englewood Cliffs, N.J., 1970.
2. A. Many, Y. Goldstein, and N. B. Grover: *Semiconductor Surfaces*. North-Holland Publishing Company, Amsterdam, 1965.
3. J. M. Ziman: *Principles of the Theory of Solids*. Cambridge University Press, New York, 1965.
4. C. Kittel: *Introduction to Solid State Physics*. John Wiley & Sons, Inc.. New York, 1966.
5. R. H. Bube: *Photoconductivity of Solids*. John Wiley & Sons, Inc., New York, 1960.

6. N. B. Hannay: *Solid-State Chemistry.* Prentice-Hall, Inc., Englewood Cliffs, N.J., 1969.

7. N. B. Hannay: in *Semiconductors,* N. B. Hannay, ed. Van Nostrand Reinhold Company, New York, 1960.

8. E. Spenke: *Electronic Semiconductors.* McGraw-Hill Book Company, New York, 1958.

9. R. A. Levy: *Principles of Solid State Physics.* Academic Press, Inc., New York, 1968.

10. P. G. Shewmon: *Diffusion in Solids.* McGraw-Hill Book Company, New York, 1963.

11. C. A. Wert and R. M. Thompson: *Physics of Solids.* McGraw-Hill Book Company, New York, 1964.

12. A. Einstein: *Ann. Physik,* **17,** 549 (1905).

13. J. M. Ziman: *Electrons and Phonons.* Cambridge University Press, New York, 1960.

14. J. C. Stone: *Radiation and Optics.* McGraw-Hill Book Company, New York, 1963.

15. R. D. Heidenreich: *Fundamentals of Transmission Electron Microscopy.* John Wiley & Sons, Inc. (Interscience Division), New York, 1964.

16. R. H. Fowler and L. W. Nordheim: *Proc. Roy. Soc. (London),* **A119,** 173 (1928).

17. R. O. Jenkins and W. G. Trodden: *Electron and Ion Emission from Solids.* Dover Publications, Inc., New York, 1967.

18. C. Herring and M. H. Nichols: *Rev. Modern Phys.,* **21,** 185 (1949).

19. J. C. Riviere: in *Solid State Surface Science,* M. Green ed. Marcel Dekker, Inc., New York, 1970.

20. J. C. P. Mignolet: *Discussions Faraday Soc.,* **8,** 326 (1950).

21. F. C. Tompkins: in *Gas-Surface Interactions,* E. A. Flood ed. Marcel Dekker, Inc., New York, 1967.

22. R. Gomer: *Field Emission and Field Ionization.* Harvard University Press, Cambridge, Mass., 1961.

23. M. Kaminsky: *Atomic and Ionic Impact Phenomena on Metal Surfaces.* Academic Press, Inc., New York, 1965.

24. M. D. Scheer and J. Fine: *J. Chem. Phys.* **46,** 3998 (1967).

25. E. W. Müller: *Advan. Electron. Electron Phys.,* **13,** 18 (1960).

26a. E. W. Müller: *Science,* **149,** 591 (1965).

26b. E. W. Müller and T. T. Tsong: *Field Ion Microscopy.* American Elsevier Publishing Company, Inc., New York, 1969.

27. H. D. Beckey: in *Mass Spectrometry,* R. J. Reed, ed. Academic Press, Inc., New York, 1965, p. 93.

28. H. Reather: *J. Phys. Radium,* **31,** C1-59 (1970).

29. K. Siegbahn et al.: *ESCA-Atomic, Molecular and Solid State Structure Studied by Means of Electron Spectroscopy.* Almqvist and Wiksells, Uppsala, 1967.

30. G. A. Somorjai and F. Szalkowski: *Advan. High Temp. Chem.*, **4** (in press, 1971).

31. O. D. Protopopov et al.: *Soviet Phys.—Solid State (English Transl.)*, **8**, 909 (1966).

Problems

4.1. The resistivity of a cadmium sulfide thin film (10^3 Å thick) changes by three orders of magnitude upon the adsorption of oxygen, from 10 ohm cm to 10^4 ohm cm. It can be assumed that the space-charge layer extends throughout the thickness of the *n*-type film and that the only electron donors are sulfur vacancies. Assuming an electron mobility of $\mu_e = 10^3$ cm²/V sec, calculate the height of the space-charge barrier at the surface, V_s.

4.2. Calculate the electron current density leaving a barium cathode under optimal conditions of emission (large accelerating potential, absence of space charge) at 600°, 800°, and 900°C. Compute the electron flux from a barium cathode surface at 800°C which has been "poisoned" by oxidation that causes a 1-V increase in work function.

4.3. Compute the degree of negative surface ionization of fluorine and oxygen at a molybdenum surface at 1800°K. Estimate the fraction of total negative charge flux emitted from the hot surface that is due to electrons assuming $j_0 = 10^{13}$ cm⁻² sec⁻¹.

4.4. The ion current leaving from the proximity of a field-ion emitter is given by $I' = I_0[k_i/(k_i + k_d)]$, where k_i is the rate constant for ionization and k_d is the rate constant for diffusion of neutral gas atoms away from the tip (R. Gomer, *Field Emission and Field Ionization*, Harvard University Press, Cambridge, Mass., 1961, p. 80). This equation reduces to $I' = I_0$ at high field strength ($k_i \gg k_d$). The ion current at 4°K was found to be about 2 per cent of the ion current at 20°K. Give reasons for the marked temperature dependence of I. How would the temperature dependence of I change with increasing applied field?

4.5. Dominant Auger peaks appear at 31, 48, and 474 eV for vanadium, 270 eV for carbon, and 150 eV for sulfur in studies of various surfaces. Using atomic tables (for example, ref. 28), list the binding energies of core electrons and assign the Auger transitions that are the most likely source of these peaks.

4.6. Semiconductor devices in thin film form are commonly used in electronic circuitries of many types. The electrical properties of these thin-film components are very sensitive to changes of the ambient conditions (moisture, oxidation, etc.), which affect the surface conductance and the space charge, owing to their high surface/volume ratio. Such thin-film circuit elements are "passivated" by the deposition of an oxide or a polymer layer that stabilizes their electrical properties. Review the various methods of passivation of semiconductor surfaces and suggest other likely methods of applying protective coating to these surfaces.

5

Interactions of Gases with Surface

5.1. Introduction

EXPERIMENTS CARRIED OUT on the surface of the earth must always take into account the possible effect of ambient gases on the experimental variables. Surface studies are particularly sensitive to ambient contamination. Thus surface chemists are continually faced with the problem of cleaning their surfaces carefully before starting an experiment. The preparation of clean surfaces is facilitated by working in high vacuum or by using freshly prepared interfaces (by cleavage, for example). Nevertheless, virtually any investigation in surface chemistry also involves the consideration of the interaction of gas molecules with the surface of the condensed phase.

The flux of gas molecules, F, which strike the surface per square centimeter per second is proportional to the gas pressure and from the kinetic theory of gases is given by

$$F\left(\frac{\text{molecules}}{\text{cm}^2\ \text{sec}}\right) = \frac{N_A P}{\sqrt{2\pi MRT}} = 3.52 \times 10^{22}\frac{P_{\text{torr}}}{\sqrt{MT}} \tag{5.1}$$

Thus at atmospheric pressures and room temperature the fluxes of gases of different molecular weights M are in the range of 10^{23} molecules/cm² sec. At $P = 10^{-6}$ torr, $F \approx 10^{14}$ molecules/cm² sec. Since it takes about 10^{15} molecules/cm² to cover a surface with a unimolecular layer, at this low pressure, the surface will be covered by gas in a few seconds if every incident molecule "sticks" to the surface. Thus one needs ambient pressures on the order of $\leqslant 10^{-8}$ torr to study the interaction of gases with initially gas-free surfaces.

The experimental measurements used in studies of the elementary steps of gas-surface interactions can be divided into two classes: (1) those which provide information about the energy transfer in gas–solid inter-actions on an atomic scale and (2) those which monitor the overall adsorption process and the properties of the adsorbed layer to yield macroscopic experimental parameters. For example, energy-transfer information is obtained from atomic and molecular beam-scattering studies (first class), while measurements that yield the gas coverage as a function of pressure (adsorption isotherms) and temperature, and heats of adsorption, belong to the second class.

In our discussion of gas–surface interactions we will first review the different molecular forces that can act between the surface and either a gas atom or molecule approaching the surface. Then we will arbitrarily divide gas–surface interactions into two types—weak and strong interactions—and discuss them separately. We will also review the fundamental energy-transfer processes that occur when a gas atom or molecule strikes the surface. We will discuss the nature of adsorption for both weakly and strongly interacting gas–surface systems. Weak inter-action leads to physical adsorption and strong interaction leads to chemisorption and frequently to chemical surface reactions. Finally we discuss the structure and the thermodynamic and electrical properties of the adsorbed layer.

5.2. Surface Forces

A gas atom or molecule "feels" an attractive potential upon approaching surfaces of different kinds. The "strength" of this potential determines the nature of the interaction between the gas and the surface atoms. The strength of the interaction can be expressed in terms of experimentally measurable parameters in several ways. The heat of adsorption ΔH_{ads} is the average binding energy per mole between the interacting gas atoms and surface atoms. ΔH_{ads}, which is related to the depth of the potential-energy well, can be obtained from adsorption or desorption studies. For $\Delta H_{ads} \gg RT_g$, where RT_g is the average thermal energy associated with 1 mole of gas atoms, the atoms undergo strong interaction with the surface, which results in partial or total energy transfer and adsorption. For $\Delta H_{ads} \approx RT_g$ the adsorption probability for incident gas atoms is small. However, if this condition is obtained by raising the surface temperature $(T_g \gg 300°K)$ to produce "hot" beams of high thermal energy to match the large interaction energy, strong interaction and complete energy transfer can still take place between the gas and the surface. This is indicated by the ease of surface ionization of alkali metals or the dissociation of hydrogen molecules on hot metal surfaces. If the condition $\Delta H_{ads} \approx RT_g$ is obtained on account of the shallow well depth of the interaction potential, the gas atoms may scatter without any appreciable energy exchange with surface atoms. Such weak inter-actions are often studied separately from strong interactions because of experimental convenience and also because of the type of assumptions that can be used to describe weak interactions (assumptions that are not at all valid for strong interactions). For convenience we will also follow this arbitrary classification in describing gas–surface interactions even though there is a gradual transition from strong to weak interactions, depending on the relative magnitudes of ΔH_{ads} and RT_g for any given gas–surface system. Adsorption studies for weakly interacting systems have to be carried out at low temperatures $(T \ll 300°K)$ to obtain measurable surface coverages necessary for these investigations. The temperature range of study distinguishes experimentally weak physical adsorption from strong chemisorption. For chemisorption, large surface concentrations of the adsorbed gas can be secured even at temperatures $T > 300°K$, because of the large attractive interaction potential between the surface and gas atoms.

Once the gas atom is adsorbed, the ratio of the well depth of the inter-action potential and the thermal energy of the gas atoms at the surface, $\Delta H_{ads}/RT$, also determines the residence time of the gas atom on the surface. The residence time τ may be estimated by

$$\tau = \tau_0 e^{\Delta H_{ads}/RT} \tag{5.2}$$

where τ_0 is related to the period of a single surface-atom vibration. For a mobile[1] adsorbed atom $\tau_0 = 1/\nu_0 \approx 10^{-12}$ sec. Measurement of the residence time can also give one an estimate of the type of interaction that takes place between the gas and the surface. Weak interaction implies residence times on the order of magnitude of τ_0, but for strong interactions $\tau \gg \tau_0$.

Frequently, the "range" of attraction of the interaction potential between the gas and surface atoms is used to describe the nature of the interaction. If the potential energy of attraction is short range, i.e., it decreases inversely as some large power of the distance between the gas and the surface atoms, it generally results in weak interaction. There are gas–surface interactions, however, characterized by electrostatic potential energy that vary inversely with the first power of the gas–surface distance and usually involve direct charge transfer or charge sharing between the colliding gas atom and the surface atom. Such an interaction potential is long range and leads to strong interaction. Experimental determination of the range of the gas–surface interaction is difficult and has been carried out only for a few systems.[2] There are several theoretical models, however, that permit one to estimate both the range and magnitude of gas–surface interactions of different types. We will describe some of these models since they provide insight into the nature of the forces exerted by the surface on the approaching gas atom or molecule.

In computing the interaction energy between the surface and the gas, the surface may be treated as a single conductor system in which the conduction electrons constitute a mobile fluctuating electron gas. As the gas atom or molecule, which has no permanent dipole moment, approaches the surface, the surface charge induces a dipole that provides the attractive force pulling the atom toward the surface. The interaction between the surface charge (which is due to the electron gas) and the gas is not much different if the gas molecule possesses a permanent dipole. In both cases the interaction energy is of the form

$$V_{L-J} = -\frac{C}{r^3} \tag{5.3}$$

where C is a constant. According to the model by Lennard-Jones[3] for spherically symmetrical atoms, C is given by $C = mc^2\chi/N_A$, where m is the electronic mass, c the velocity of light, N_A Avogadro's number, and χ the diamagnetic susceptibility of the gas atom. C is on the order of 10^2 kcal/mole when r is given in angstroms. In Table 5.1 the surface-interaction energies of several monatomic and diatomic gases are listed at a distance of closest approach of 4 Å. The interaction potential between metal surfaces and approaching gas atoms has the same range as given in Eq. (5.3) as was shown by Bardeen[4] and by Margenau and Pollard,[5] and the constant of proportionality is of the same magnitude as that derived by Lennard-Jones. It should be noted that in using these

Table 5.1. *Values of the Constants of the Lennard-Jones and London Interaction Potentials and the Interaction Energies at the Distance of Closest Approach of* 4 Å

	C (kcal Å^3/mole)	C'(kcal Å^6/mole)	V_{L-J} (kcal/mole) $(r = 4$ Å)	V_{London} (kcal/mole) $(r = 4$ Å)
Ne	129	67	2.0	0.016
Ar	352	802	5.5	0.19
H$_2$	76	176	1.2	0.04
N$_2$	144	919	2.2	0.22
CO$_2$	362	1872	5.6	0.46

models of gas–surface interactions the gas atom is assumed to interact with the surface *as a whole* instead of individual surface atoms. Recently[6] there has been experimental evidence that in some cases the interaction potential between a metal surface and organic molecules of different types varies inversely as the square of the distance, $V \propto r^{-2}$.

For gas–surface interactions of certain types it may be useful to view the interaction as between the gas atom and a *single* surface atom. Weak attractive interaction between a pair of atoms can be due to dispersion forces (London[7]) that represent the interaction of induced fluctuating charge distributions. In addition, molecules that possess permanent dipoles can further polarize each other (Debye[8]) and can have dipole–dipole interactions (Keesom[9]). All these pairwise interaction potentials fall off inversely as the sixth power of the distance.

The dispersion force is due to induced dipole interaction between atoms or molecules through electron density fluctuations. The potential energy of interaction, V_{London}, is given by

$$V_{London} = -\frac{C'}{r^6} \tag{5.4}$$

where, using an approximate model, C' is given by

$$C' = \frac{3}{2} \frac{h\nu_1\nu_2}{\nu_1 + \nu_2} \alpha_1\alpha_2$$

Here α_1 and α_2 are the polarizabilities of the interacting species, ν_1 and ν_2 are their characteristic frequencies of oscillation (oscillator strengths), and h is Planck's constant. C' can be calculated to be on the order of 10^3 kcal Å^6.

Since the values of ν_1 and ν_2 are not easily available, one seeks other ways to estimate the dispersion constant C' from readily measurable molecular properties. One very good approximate expression,[10] which was developed by Slater and Kirkwood, can be written as

$$C' \left(\frac{\text{kcal } \text{Å}^6}{\text{mole}} \right) = 363\alpha_1\alpha_2 \Big/ \left[\left(\frac{\alpha_1}{n_1}\right)^{1/2} + \left(\frac{\alpha_2}{n_2}\right)^{1/2} \right] \tag{5.5}$$

where α_1 and α_2 are the polarizabilities in units of Å^3, and n_1 and n_2 are the number of electrons in the outer shells of the molecules. In Table 5.1 the London interaction energies are also listed for a distance of $r = 4 \text{ Å}$ along with the dispersion constants for several pairs of like atoms (for interaction between like species, $\alpha_1 = \alpha_2$ and $n_1 = n_2$). It can be seen that due to its short range, V_{London} is a much weaker attractive potential at that distance when compared with the $V \propto r^{-3}$ type of attractive potential.

Molecules that possess permanent dipole moment can further polarize each other, which gives, for the mutual attractive potential energy V_{Debye},

$$V_{\text{Debye}} = -\frac{\alpha_1\mu_2^2 + \alpha_2\mu_1^2}{r^6} \tag{5.6}$$

where μ_1 and μ_2 are the dipole moments of the interacting molecules. Direct interaction of two different molecules with permanent dipoles without additional polarization yields

$$V_{\text{Keesom}} = -\frac{2}{3k_B T} \frac{\mu_1^2\mu_2^2}{r^6} \tag{5.7}$$

Table 5.2. *Average Polarizabilities for Several Atoms and Molecules*

	$\bar{\alpha}$ (Å^3)		α (Å^3)
Neon (Ne)	0.39	Hydrogen sulfide (H_2S)	3.78
Argon (Ar)	1.63	Ammonia (NH_3)	2.26
Krypton (Kr)	2.46	Nitrous oxide (N_2O)	3.00
Xenon (Xe)	4.00	Methane (CH_4)	2.60
Hydrogen (H_2)	0.79	Ethane (C_2H_6)	4.47
Nitrogen (N_2)	1.76	Ethylene (C_2H_4)	4.26
Oxygen (O_2)	1.60	Benzene (C_6H_6)	10.32
Carbon monoxide (CO)	1.95	Acetone (CH_3COCH_3)	6.33
Carbon dioxide (CO_2)	2.65		

Table 5.3. *Dipole Moments of Several Molecules*

Substance	μ (Debye)
H_2O	1.84
H_2S	0.89
NO	0.16
CO	0.12
N_2O	0.166
HF	1.91
HCl	1.08
NH_3	1.45
CH_3OH	1.68
CH_3CHO	2.72
$(CH_3)_2CO$	2.9

1 Debye $= 1 \times 10^{-18}$ esu

Both V_{Debye} and V_{Keesom} are orientation averaged expressions and Eq. (5.7) is restricted to gases in thermal equilibrium. The dispersion interaction V_{London} is appreciably larger than these other two effects (except for the most polar molecules, such as water, for which V_{Keesom} is somewhat larger). Table 5.2 lists the average polarizabilities of several atoms and molecules; Table 5.3 lists the dipole moments of several molecules. There are also many other types of interactions (ion-induced dipole that varies as r^{-4}, etc.), but they are likely to be less important in gas–surface interactions and will not be discussed here.

It has been shown[11] that the dispersion interaction between pairs of atoms is additive. Calculations show[12] that a large long-range attractive interaction may result from the simultaneous dispersion interaction of many atoms. For example, the attractive potential energy of interaction between two flat plates, V_{plate}, in vacuum, due to the summation of the pairwise dispersion forces, varies inversely with the square of the distance, $V \propto r^{-2}$.

There are atoms and molecules that ionize or participate in partial charge transfer when colliding with surface atoms. In this case the interaction energy is very strong and long-range and may be represented by a Coulomb potential

$$V = -\frac{e_1 e_2}{r} \tag{5.8}$$

where e_1 and e_2 are the charges associated with the gas and with the surface atom, respectively. In this case the interaction energy varies inversely with the first power of the distance.

The magnitude of the strong interaction energy that leads to surface ionization is determined by the work function of the solid, ϕ, and the ionization potential of the approaching molecules. The interaction energy between cesium and sodium atoms and tungsten surfaces is on the order of 65 and 28 kcal/mole, respectively, much larger than that which would be due to solely dispersion forces. The interactions of oxygen and hydrogen with metal surfaces which involve charge transfer but do not lead to complete surface ionization are additional examples of strong surface forces and result in interaction energies in the range 20–200 kcal/mole. Thus gas–surface interaction energies can be as large as the lattice energies that hold the condensed phases together.

Gas–surface interactions of this type lead to the formation of surface chemical bonds. The atoms chemisorbed in such a manner may often participate in subsequent surface reactions.

5.3. Atomic Scattering Studies of Weak Interactions of Gas Atoms with Surfaces

In this section we will discuss some of the fundamental energy-transfer processes that occur when a gas atom strikes a surface. These are studied by the powerful technique of atomic or molecular beam scattering. In these experiments a "beam" of gas atoms that is monoenergetic or has a well-defined velocity distribution is allowed to strike a surface (preferably a single crystal with known temperature, orientation, roughness, etc.) at a given angle of incidence. The flux, velocity, and/or the density of the scattered beam is detected as a function of scattering angle (angular distribution) by using a detector (usually a mass spectrometer or ion gauge). From the properties of the incident beam (velocity, direction, etc.), the surface (temperature, atomic weight, orientation, etc.), and the scattered beam (velocity, angular distribution, etc.), the dynamics of the gas–surface interaction can be determined.

Different phenomena may occur during the collision of gas atoms with surface atoms: (1) the gas atoms may diffract from the surface (elastic scattering) or (2) the gas atoms may exchange energy with the surface (inelastic scattering). Energy transfer between the gas and surface atoms occurs most frequently in collision processes. The nature and the extent of the energy transfer depends on the velocity and mass of the incident atom, on ΔH_{ads}, its residence time near the surface, and on the surface characteristics (surface temperature, mass of surface atoms, surface roughness, etc.) that determine the lattice vibration spectra of surface atoms.

The inelastic scattering processes may be divided into two classes. The gases that undergo weak interaction with the surface atoms belong to the first class. The gas atoms reside near the surface for a time on the order of 10^{-12} sec (which is the same magnitude as the period of lattice vibrations for surface atoms) before being scattered. Because of this short residence time, there is, in general, a correlation between the directions of the incident and scattered beams and their velocities. By using simple models for the weak interaction between the gas atoms and the surface atoms, the velocity, flux, and density distribution of the scattered atoms can be calculated as a function of the experimental parameters that characterize the incident beam and the surface. Calculations are given below of the scattered beam properties based on one of these simple models giving good agreement with experiments. The second class of inelastic scattering processes characteristic of strongly interacting gas–surface systems will be discussed in Section 5.5.

5.3.1. DIFFRACTION OF GAS ATOMS BY SURFACES

The average velocity of atoms $\bar{c}$ in a volume of gas at temperature T, and the root-mean-square velocity $(\bar{u}^2)^{1/2}$ is given by $\bar{c} = (8k_B T/\pi M)^{1/2}$ and $(\bar{u}^2)^{1/2} = (3k_B T/M)^{1/2}$, respectively, where M is the mass of the gas atom and the other constants have their usual meaning. It is assumed that the gas has reached thermal equilibrium with the walls of the source so that the gas atoms can be characterized by a "gas temperature" T_g that is identical with the source temperature $(T = T_g)$. The average kinetic energy of the gas atoms, $\bar{T}$, due to their three translational degrees of freedom is given by

$$\bar{T} = \tfrac{1}{2}M\bar{u}^2 = \frac{3k_B T_g}{2} \tag{5.9}$$

Thus a stream of helium atoms that effuses from a source held at room temperature ($T = 300°K$) has a root-mean-square velocity $(\bar{u}^2)^{1/2} \approx 10^5$ cm/sec and a translational energy of $\bar{T} \approx 0.04$ eV. We can calculate the wavelength associated with the atoms by using the deBroglie equation,

$$\lambda = \frac{h}{(2M\bar{T})^{1/2}} \tag{5.10}$$

as indicated in Chapter 1. For a thermal helium atom $\lambda \approx 1$ Å, so diffraction of these gas atoms from a row of surface atoms is possible.

By suitable choice of source geometry and by using "velocity selection" it is possible to produce a unidirectional beam of atoms with well-defined energy. There are many different designs of velocity selectors; perhaps the simplest consists of two (or more) slotted rotating discs (Fig. 5.1).

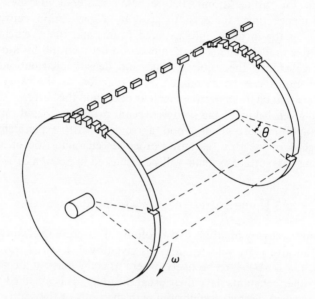

Figure 5.1. *Velocity selector for molecular-beam studies.*

The discs rotate with an angular velocity which is of the same order of magnitude as the desired beam velocity.[13] The transmitted velocity u is

$$u = \frac{nl\omega}{2\pi} \tag{5.11}$$

where l is the distance between the discs, ω (in rad sec^{-1}) is the angular frequency, and n is the number of slits in each disc. Since the time of flight t of the atoms between the rotating discs is $t = l/u$, suitable displacement of the slots of the two discs with respect to each other ensures that atoms only in a well-defined velocity range are transmitted and are not being scattered by the second rotating disc. The use of additional selector discs eliminates the transmission of higher harmonics of the velocity u (i.e., $2u$, $3u$, etc.).

The atomic beam, after scattering from a surface, may be detected by a mass spectrometer, ion gauge, or a surface-ionizer detector. The angular distribution of the scattered beam is measured by rotation of the detector about the fixed sample or by rotation of the sample and beam source when using a fixed detector. The angle of incidence can also be varied by using suitable experimental geometry. The scheme of one type of molecular-beam-scattering apparatus is shown in Fig. 5.2.

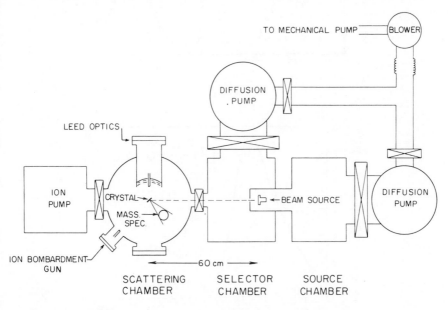

Figure 5.2. *Schematic diagram of molecular-beam–surface scattering system.*

Diffraction of helium from cleaved (100) surfaces of lithium fluoride crystals was detected by Stern and his coworkers in 1929,[14a] and their experiment has been repeated and verified more recently.[14d] The experimental curves obtained are shown in Fig. 5.3(a) and (b). The scattered beam intensity is plotted as a function of scattering angle, clearly indicating the presence of the specular (00) and first-order diffraction beams (0 ± 1). LiF acted as a two-dimensional grating of one set of ions only and the peaks appeared at

$$\cos \alpha_r - \cos \alpha_i = \frac{m}{a} \lambda \tag{5.12}$$

$$\cos \beta_r - \cos \beta_i = \frac{n}{a} \lambda \tag{5.12a}$$

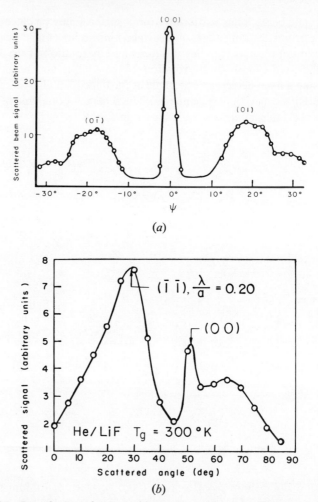

Figure 5.3. *Distribution of scattered helium beam from the (100) face of lithium fluoride displaying the (a) (00), ($0\bar{1}$), and (01) and (b) (00), ($\bar{1}\bar{1}$) diffraction peaks.*

where $\cos \alpha$ (defined in Fig. 5.4) and $\cos \beta$ are the direction cosines and a is the shortest distance between two ions of like charges.

Diffraction of helium beams from metallic (silver) surfaces has also been attempted but so far without success. Since clean and well-ordered surfaces are necessary to observe diffraction, experimental difficulties may have been the cause of the lack of success. It has been proposed that the large free electron density at metal surfaces (Chapter 4) and the large Debye–Waller factor for surface atoms (Chapter 2) modify the surface potential "felt" by the incident helium atom so as to minimize

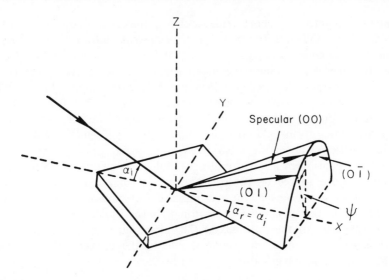

Figure 5.4. *Definition of scattering angles used to identify the helium diffraction beams.*

its periodic nature. Clearly, more diffraction experiments have to be carried out on a variety of solid surfaces to further explore the nature of atomic beam diffraction.

5.3.2. WEAK INELASTIC SCATTERING OF GAS ATOMS FROM SURFACE

During the scattering of the incident atomic beam the surface atoms responsible for the scattering do not remain stationary but vibrate about their equilibrium positions. The average energy $\bar{E}$ associated with this lattice vibration at higher temperature $[T_s > \Theta_D \text{ (surface)}]$ can be estimated as $\bar{E} \approx 3RT_s$, where T_s is the surface temperature. For $T_s = 300°\text{K}$, $\bar{E} \approx 0.078$ eV. Thus the energy associated with lattice vibrations is of the same magnitude as the kinetic energy of the incident atomic beam. Therefore, direct energy transfer between the incident beam and the surface can take place, which could markedly alter the angular and energy distribution of the scattered atomic beam. Indeed, a large fraction of the incident atoms undergoes inelastic scattering with the surface. If $T_s < T_g$, part of the kinetic energy of the gas atom is more likely to be transferred to the surface during the collision. For $T_s > T_g$, energy transfer from the surface to the gas atom is more probable.

Weak interactions between surface and gas atoms are characterized by small values of the heat of adsorption, ΔH_{ads} (shallow well depth for the interaction potential), short residence time of the gas atom on the surface, and a short-range interaction potential. The weak collisional interaction can thus be considered as similar to that between two billiard balls or hard rubber balls, where the momentum exchange between the two colliding particles is predetermined by the incident angle, mass, and velocity of the two particles. Interactions of this type result from surface scattering of rare gas atoms and the scattering of fast, high-thermal-energy atoms from cold surfaces. Atomic and molecular beams with kinetic energies in the range 0.5–10 eV ($RT_g > \Delta H_{ads}$) can be produced by the expansion of a mixture of light and heavy gas molecules from a high-pressure source ($\sim 10^3$ torr) through a nozzle or small orifice. The heavy gas atoms in the gas mixture will have a kinetic energy, $\frac{1}{2}m(\text{heavy})u^2$ that is larger by a factor of $[m(\text{heavy})/m(\text{mean})]$ than the kinetic energy of the pure gas at the same source temperature. Here $m(\text{mean})$ is the mean molecular weight in the mixture. The kinetic energy of the heavy gas atoms is given by the equation[15]

$$\tfrac{1}{2}m(\text{heavy})u^2 = \frac{m(\text{heavy})}{m(\text{mean})} \int_0^{T^{\text{source}}} C_P(\text{mean})\, dT$$

There have been several calculations to describe the dynamics of weak gas–surface collisions. We will discuss only one of these; it is based on the "hard-cube" scattering model,[16,17] which leads to a closed-form expression of the angular distribution of the scattered beam.[18] The results of the calculations with this model can then be directly correlated with the results of rare-gas atomic-beam-scattering studies from various surfaces. In the model for weak interactions, the gas and surface atoms are assumed to be rigid elastic particles. The surface is assumed to be perfectly smooth, and the surface atoms are represented by small blocks that can move only normal to the surface and which are confined by independent square-well potentials. The extent of the momentum transfer $\Delta(Mu)$ in the gas–surface collision can be investigated by monitoring the angular distribution of the scattered beam at a given angle of incidence. The change of angle reflects the change of momentum of the gas atom during collision with the surface. In the hard-cube scattering model *the tangential velocity of the gas atom* (i.e., velocity component along the surface) *is assumed unchanged by the collision* because no forces act parallel to the surface. Furthermore, the gas–solid interaction potential is assumed to be uniform across the plane of the surface and weak enough to permit impulsive collisions [i.e., the gas atom has a very short residence time on the surface ($\sim 10^{-12}$ sec)]. It is thus convenient to think of the surface atoms as being represented by small cubes oriented with one face parallel to the surface and moving only normal to the surface. Each gas particle interacts with only one of these cubes during the collision. The total scattering cross section of the cube is thus equal to the area $\mathscr{A}$ of its exposed face. While the tangential velocity of the incident gas particle is unchanged, the classical laws governing collisions between rigid elastic bodies will be used to calculate the normal component of the velocity for the scattered particles. This classical model of gas–surface collisions is shown schematically in Fig. 5.5. A gas atom of

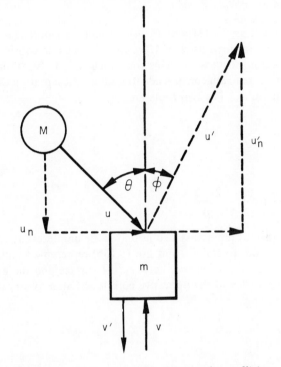

Figure 5.5. *Hard-cube model of gas–surface collisions.*

mass M and speed u is incident upon a surface atom of mass m that can move with velocity v, only normal to the surface plane. The angle of incidence, θ, is measured from the surface normal. The gas atom is scattered with speed u' and at an angle ϕ, which is also measured from the surface normal.

Conservation of energy before and after the collision requires that

$$\tfrac{1}{2}Mu_n^2 + \tfrac{1}{2}mv^2 = \tfrac{1}{2}Mu_n'^2 + \tfrac{1}{2}mv'^2 \tag{5.13}$$

where we have introduced the speed v' (presumed downward) of the surface atom after the collision and u_n and u_n' are the normal components of u and u', respectively. Since it is assumed that the tangential velocity component of the gas particle remains unchanged in the collision and that it recoils instantly ($\tau \approx \tau_0$), the energy of the scattered atom can be deduced directly from the properties of the incident beam and ϕ, the scattering angle. Conservation of momentum in the direction normal to the surface yields

$$Mu_n - mv = -Mu'_n + mv' \tag{5.14}$$

By combining Eqs. (5.13) and (5.14), v' can be eliminated. We also define the mass ratio μ as $\mu = M/m$ and restrict our consideration of surface scattering to those systems for which $\mu < 1$. Substitution of μ and subsequent rearrangement leads to a quadratic equation that can be solved to obtain u'_n. The two roots are

$$u'_n = -n_n \tag{5.15a}$$

and

$$u'_n = \frac{1-\mu}{1+\mu} u_n + \frac{2}{1+\mu} v \tag{5.15b}$$

Clearly Eq. (5.15a) represents a solution for the uninteresting case where the two atoms fail to collide, and Eq. (5.15b) represents a solution that permits the exchange of energy between the surface and the gas atoms.

From the geometry of the impulsive collision (Fig. 5.5) we can express the scattering angle ϕ as

$$\cot \phi = \frac{u'_n}{u_n} \cot \theta \tag{5.16}$$

It will be useful to express the velocity of the surface atom, v, as a function of u and ϕ by combining Eqs. (5.15b) and (5.16) to eliminate u'_n as

$$v = \left(\frac{1+\mu}{2} \sin \theta \cot \phi - \frac{1-\mu}{2} \cos \theta \right) u \equiv B_1 u \tag{5.17}$$

and its differential as a function of the scattering angle, $dv/d\phi$, which is given by differentiating Eq. (5.17) as

$$\frac{dv}{d\phi} = -\left(\frac{1+\mu}{2} \sin \theta \csc^2 \phi \right) u \equiv B_2 u \tag{5.18}$$

Now that we have an expression for the velocity of the scattered beam, u', in terms of the incident velocity u and the "velocity" of the surface atom normal to the surface plane, v [Eq. (5.15b)], both of which can be determined by experiments, we should calculate the total collision rate Z between gas atoms and surface atoms. The total collision rate Z between surface atoms and gas atoms is $Z =$ (collision cross section) $\times$ (relative collision velocity) $\times$ (density of surface atoms) $\times$ (density of gas atoms)

$$Z = \mathcal{A} V_R n_s n_g \tag{5.19}$$

where n_s(atoms/cm²) and n_g(atoms/cm³) represent the density of surface and gas atoms, and V_R is the "relative" speed of collision, which is given by the sum of the normal component of the incident velocity $u_n = u \cos \theta$ and the speed of the surface atom v:

$$V_R = u \cos \theta + v \tag{5.20}$$

In order to evaluate the probability of scattering of one gas atom with a given incident angle θ into a scattering angle ϕ, one has to determine the differential collision rate $dR/d\phi$. This is carried out in two steps. First we express the differential rate at which collisions occur between incident gas atoms with velocity between u and $u + du$ and surface atoms with velocities between v and $v + dv$. Then we calculate the probability for scattering into an angular range between ϕ and $\phi + d\phi$ by writing $v = v(u, \phi)$ and integrating over all incident velocities. The differential rate of collision between gas and surface atoms is given by

$$\frac{d^2 R}{du\, dv} = Z F(u) G(v) \tag{5.21}$$

where $F(u)$ is the normalized velocity distribution of the incident beam and $G(v)$ is the normalized velocity distribution for the surface atoms. The differential rate of collision *per surface atom* is obtained by dividing both sides of Eq. (5.21) by n_s. Substitution of Eq. (5.19) into (5.21) yields

$$\frac{1}{n_s} \frac{d^2 R}{du\, dv} = \mathcal{A} V_R n_g F(u) G(v) \tag{5.22}$$

Now we may change variables in Eq. (5.22), assuming that $v = v(u, \phi)$ so that $dv = (\partial v / \partial \phi)_u\, d\phi$, to obtain

$$\frac{1}{n_s} \frac{d^2 R}{du\, d\phi} = \mathcal{A} V_R n_g F(u) G(v) \left(\frac{\partial v}{\partial \phi}\right)_u \tag{5.23}$$

One can now integrate over all values of the incident velocity u. In order to obtain the probability distribution of scattered atoms $P(\phi)$ we must divide the integrated form of Eq. (5.23) by the number of atoms striking the surface per unit time $\mathcal{A}n_g\bar{u}_n$ ($\bar{u}_n$ is the average normal velocity):

$$P(\phi) = \frac{1}{\mathcal{A}n_s n_g \bar{u}_n} \frac{dR}{d\phi} = \frac{1}{\bar{u}_n} \int_{u=0}^{\infty} V_R F(u)G(v)\left(\frac{\partial v}{\partial \phi}\right)_u du \qquad (5.24)$$

Substitution of Eqs. (5.17), (5.18), and (5.20) into (5.24) yields

$$P(\phi) = \frac{1}{\bar{u}_n} \int_{u=0}^{\infty} (\cos\theta + B_1)B_2 u^2 F(u)G(B_1 u_n) \, du \qquad (5.25)$$

Let us assume that the gas atoms have a three-dimensional Maxwellian distribution[19]

$$F(u)\, du = 4\pi u^2 \left(\frac{M}{2\pi k_B T_g}\right)^{3/2} \exp\left(-\frac{Mu^2}{2k_B T_g}\right) du \qquad (5.26)$$

and let us also assume that the surface atoms whose motion is restricted to one dimension and perpendicular to the surface plane have a one-dimensional Maxwellian distribution[19]

$$G(v)\, dv = \left(\frac{m}{2\pi k_B T_s}\right)^{1/2} \exp\left(-\frac{mv^2}{2k_B T_s}\right) dv \qquad (5.27)$$

where T_s is the surface temperature. Substitution of Eqs. (5.26) and (5.27) into (5.25), which is then integrated [using the integral $\int_0^\infty x^4 \exp(-ax^2)\, dx = (3/8a^2)(\pi/a)^{1/2}$], yields

$$P(\phi) = \frac{3}{4}\left(\frac{mT_g}{MT_s}\right)^{1/2} B_2(1 + B_1 \sec\theta)\left(1 + \frac{mT_g}{MT_s}B_1^2\right)^{-5/2} \qquad (5.28)$$

This equation gives the scattering probability of a flux of gas atoms into a unit angular range $d\phi$ at angle ϕ. It appears that the angular distribution depends only on (1) the mass ratio $\mu = M/m$, which is the ratio of masses of gas and surface atoms, (2) the ratio of surface temperature to characteristic gas temperature T_s/T_g, and (3) the incident angle θ. It should be noted that the ratio, $(MT_s/mT_g)^{1/2}$, is the ratio of the mean speed of the surface atom to that of the gas atom. The angular distribution of scattered atoms seems to be a strong function of this ratio. Figure 5.6 shows the experimental data for helium scattered from the

(100) face of platinum. The scattered distribution is shown in both polar and rectilinear plots that are used interchangeably in the literature.

For scattering of monoenergetic atomic beams, a single velocity u can be used and substituted into Eq. (5.25) instead of the Maxwellian distribution $F(u)$, as given in Eq. (5.26). Under these conditions the integration of Eq. (5.25) yields the simple expression

$$P(\phi) = \frac{B_2}{\sqrt{\pi}} u \left(\frac{m}{2k_BT_s}\right)^{1/2} (1 + B_1 \sec \theta) \exp \left(-\frac{B_1^2mu^2}{2k_BT_s}\right) \qquad (5.29)$$

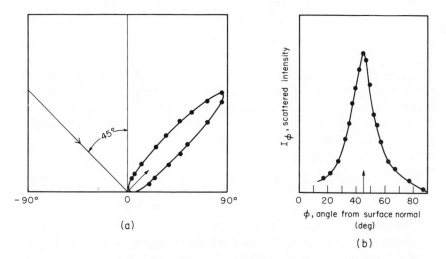

(a)

(b)

Figure 5.6. *Distribution of scattered helium beams from the (100) face of platinum single crystal displayed on a (a) polar and on a (b) rectilinear plot.*

It is important to reiterate the major assumptions of the scattering model that yielded Eq. (5.28): There are only single collisions between gas and surface atoms, and these are considered to be impulsive (rigid, elastic particles); collisions with surface atoms did not change the tangential velocity component of the gas atom; the surface is smooth and the surface atoms are represented by independent one-dimensional oscillators that have Maxwellian velocity distributions. Thus this model, which represents well the scattering of rare gas atoms of thermal energy from smooth metal surfaces with $\mu \leqslant 0.3$, will likely break down if (1) multiple collisions cannot be neglected, (2) the well depth of the surface potential is large compared to the energy of the incident particle $\Delta H_{ads} > RT_g$, and (3) internal degrees of freedom are affected by the collision. In short, under the conditions of strong interactions between the gas and the surface, which lead to increased residence time, and marked energy transfer between the collision partners this weak-interaction model is not expected to hold.

Frequently it is not the angular distribution of scattered atoms which is given by Eq. (5.28) but their velocity distribution that is of interest. This is the case if one studies the nature of the energy transfer between the gas and surface atoms. It can be shown, by a derivation similar to that which yielded Eq. (5.28), that the hard-cube model gives the velocity distribution of the scattered flux of atoms, $P(U)$, as

$$P(U) = \frac{8}{3\pi^{1/2}} B_3^5 \left(1 + \frac{mT_g}{MT_s} B_1^2 \right)^{5/2} U^4 \exp \left[-B_3^2 \left(1 + \frac{mT_g}{MT_s} B_1^2 \right) U^2 \right]$$

(5.30)

where B_3 is defined as $B_3 = \sin \phi / \sin \theta$, U is a dimensionless velocity defined by $U = u_n (M/2kT_g)^{1/2}$, and all other terms have been defined previously. The most probable velocity u^*, which appears at a given scattering angle ϕ in the flux, can be obtained by setting the derivative of Eq. (5.29) equal to zero. The result is

$$u^*(\phi) = \frac{\sqrt{2}}{B_3} \left(1 + \frac{mT_g}{MT_s} B_1^2 \right)^{-1/2}$$

(5.31)

In most atomic beam scattering experiments the experimental data are obtained by monitoring the scattered beam intensity I as a function of scattering angle ϕ while keeping the other variables—θ, T_g, T_s, and μ—constant. Then one of these parameters is varied (the surface temperature T_s is changed, for example) and the I–ϕ curve is redetermined. The detectors used in monitoring the scattered beam either measure the scattered density distribution n_g(atoms/cm³) [open mass spectrometer ionizer, for example] or the scattered flux distribution F(atoms/cm² sec) = $n_g \bar{c}$ [closed ionization gauges, for example]. The velocity of the scattered beam can be determined by measuring simultaneously both the density and the flux of the scattered beam since $\bar{c} = F/n_g$. Another technique for measuring the velocity of the scattered beam involves measurement of the change in its time of flight.[20] Since the distance between the surface and the detector and the incident velocity is known, the scattered beam velocity can be calculated.

As shown in Fig. 5.7, the scattered beam has a lobular distribution with a maximum at a characteristic scattering angle ϕ_{max}. The hard-cube scattering theory predicts well the following dependences of ϕ_{max} on the variables M, T_s, T_g, and θ, which were proved by experiments:

1. $\phi_{\max}$ moves toward the surface normal from the specular direction (where the incidence angle θ is equal to the scattering angle ϕ) with increasing atomic weight of the scattering gas atom (Fig. 5.7). In these experiments T_s, T_g, and θ are held constant. Thus, the perpendicular component of the momentum of the scattered beam increases with increasing M. Such a trend of increased scattering toward the surface normal is often called "subspecular" scattering.

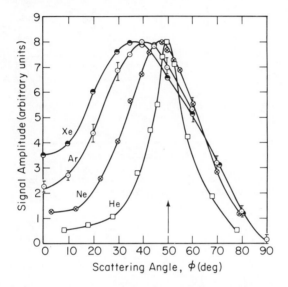

Figure 5.7. *Angular distribution of He, Ne, Ar, and Xe atomic beams scattered from oriented Ag (111) films.*

2. $\phi_{\max}$ moves toward the surface normal from the specular direction with increasing surface temperature, T_s (Fig. 5.8). In these experiments M, T_g, and θ are held constant. Such a subspecular scattering again indicates strong momentum exchange interaction between the incident gas atom and the surface atom.

3. $\phi_{\max}$ moves toward the surface plane from the specular direction with increasing characteristic gas temperature T_g (i.e., increasing gas velocity). This increase in the tangential component of the momentum indicates stronger momentum transfer to the surface with increasing T_g (Fig. 5.9). In these experiments M, T_s, and θ are held constant.

4. $\phi_{\max}$ moves toward the surface from the specular direction with increasing θ for small values of θ at nearly normal incidence. (θ is the angle of incidence as measured from the surface normal.) $\phi_{\max}$ moves toward the surface normal; i.e., it changes direction as θ approaches the

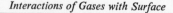

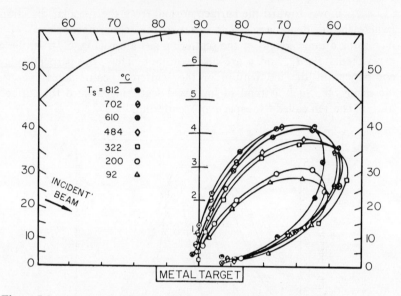

Figure 5.8. *Angular distribution of argon atomic beams scattered from platinum foil at different surface temperatures, T_S.*

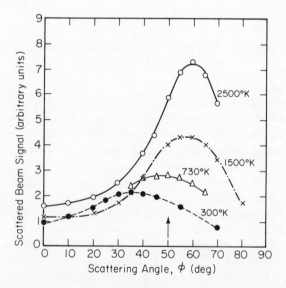

Figure 5.9. *Angular distribution of xenon atomic beams scattered from oriented Ag (111) films at various beam temperatures, T_g.*

surface. This result cannot be rationalized easily; it is apparent that θ is not independent of M, T_s, and T_g. Thus the nature of the energy transfer in scattering experiments with varying angle of incidence depends on the magnitude of the other three parameters as well. There is insufficient experimental information on the variation of ϕ with θ to allow unambiguous comparison of the theory with experiments.

During the scattering of monatomic gases by surfaces there may be direct energy transfer between surface phonons and the incident gas atoms. Therefore it should be possible to obtain information about the energy distribution of the surface phonons from the changed energy and angular distribution of the scattered atomic beam. Helium atomic beams may play the same role in obtaining surface phonon spectra as neutrons do in obtaining the spectra of bulk phonons by similar experiments.

Weak interactions between gas and surface atoms lead to minimal energy transfer between the colliding particles. In this circumstance, the number density distribution of the scattered beam as a function of angle, $n(\phi)$, is not expected to be much different from the flux distribution $F(\phi)$, which also depends on the velocity of the scattered atoms. Large differences between $n(\phi)$ and $F(\phi)$ are expected, however, in case of stronger gas–surface interactions, which lead to more complete energy exchange.

5.3.3. ACCOMMODATION COEFFICIENTS

Frequently one is more interested in determining the energy distribution of the scattered beam than in measuring its angular distribution. One would like to determine from the angular distribution of the scattered beam how the energy transfer between the incident gas atom and the surface depends on the experimental variables μ, T_s, T_g, and θ.

To this end, one may define an *energy accommodation coefficient* α_E, as

$$\alpha_E = \frac{E_i - E_r}{E_i - E_s} \tag{5.32}$$

where E_i and E_r are the kinetic energies of the incident and reflected atomic beam, respectively, and E_s is the mean energy of the gas atoms at the surface temperature, T_s. The energy accommodation coefficient so defined takes up values between 0 and 1. If there is no energy exchange between the gas and the surface, $E_i = E_r$ and $\alpha_E = 0$. For complete energy accommodation on the surface, $E_r = E_s$ and $\alpha_E = 1$. It appears that for weak interactions, i.e., scattering of rare gases from metal surfaces, energy transfer occurs primarily through the momentum component normal to the surface.[21] α_E increases with increasing M and T_g. This is the probable reason for the absence of diffraction effects in the scattering of monatomic gases other than helium (Kr, Ar, Xe). The energy accommodation coefficient decreases with increasing T_s and with increasing θ. In addition it was found that α_E decreases with increasing lattice force constant (which is proportional to the square of the effective

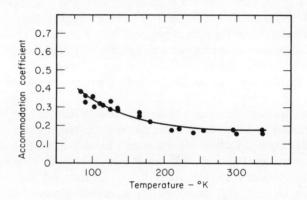

Figure 5.10. *Accommodation coefficient of argon on tungsten polycrystalline surfaces as a function of surface temperature.*

surface Debye temperature). Finally, α_E increases as the potential between the incident gas atom and the surface atom becomes more attractive.

Usually the energy exchange between the gas and the surface are studied by *thermal accommodation* experiments (rather than by the more difficult molecular-beam techniques). In these investigations one determines the thermal energy of the gas (i.e., gas-temperature change) before and after the partial energy exchange with the surface. The thermal accommodation coefficient is commonly defined as

$$\alpha(T) = \frac{T_g - T_g'}{T_g - T_s} \tag{5.33}$$

where T_s is the surface temperature and T_g and T_g' are the temperatures associated with the gas before and after scattering from the surface, respectively. Equation (5.33) is similar to the definition of the energy accommodation coefficient, Eq. (5.32). If there is no exchange of thermal energy, $T_g' = T_g$ and $\alpha(T) = 0$. If there is complete thermal accommodation, $T_g' = T_s$ and $\alpha(T) = 1$.

There are several techniques used for measuring $\alpha(T)$. For example, one measures the change in thermal conductivity of the metal filament during its contact with the gas. From these data, $\Delta T = T_g - T_g'$ can be calculated. From the additional determination of T_s and T_g during the experiment, $\alpha(T)$ is evaluated as a function of gas pressure, temperature, or other variables. In Figs. 5.10 and 5.11 recent values of the accom-

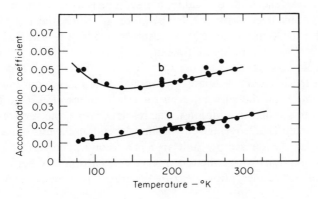

Figure 5.11. *Accommodation coefficient of (a) helium and (b) neon on tungsten polycrystalline surfaces as a function of surface temperature.*

modation coefficients of argon, helium, and neon are given for tungsten surfaces.

Macroscopic thermal accommodation studies provide a great deal of information about the experimental parameters (surface impurities, competitive adsorption of other gases, etc.) that influence the energy transfer between the gas and the surface. Nevertheless, valuable additional information could be obtained by measuring the angular distribution of the scattered beam and by using single-crystal surfaces in these investigations.

It is important to consider the effect of surface impurities on the thermal accommodation coefficient of a gas. Light adsorbed impurity atoms for which the atomic mass is smaller than that for atoms in the host lattice tend to facilitate the energy transfer between the gas and the surface, as shown by experiments.[22] Thus the adsorption of light atoms increases the thermal accommodation coefficient. The presence of heavy impurity atoms, for which the atomic mass is greater than that for atoms in the clean surface, is expected to influence the thermal accommodation coefficient to a lesser extent.

5.4. Physical Adsorption

5.4.1. CONCEPTS

We will now concern ourselves with the properties of the adsorbed layer of a weakly interacting gas on a solid surface. We can see from Eq. (5.2) that the residence time of weakly interacting gas atoms can be increased markedly by decreasing the temperature at which the experiment is carried out. Assuming a heat of adsorption $\Delta H_{ads} \approx 2$ kcal/mole, and $\tau_0 = 10^{-12}$ sec, the residence time at $300°K$ is on the order of 10^{-11} sec. At $100°K$, however, it is greater than 10^{-7} sec! Thus by judicious choice of the experimental conditions we can maintain a large concentration of gas atoms σ on the surface, even for small values of ΔH_{ads}. It is not difficult to see that in addition to the residence time, the surface concentration or surface "coverage" σ will also depend on the flux of gas atoms, F, striking the surface per unit area per second. The surface coverage under conditions where there is a large concentration of surface atoms still available as adsorption sites (low coverage), is given by the product,[23]

$$\sigma = \tau F \tag{5.34}$$

The gas flux is proportional to the pressure and from the kinetic theory of gases it is given by Eq. (5.1); the residence time is defined in Eq. (5.2). Using these two equations, we can rewrite the surface coverage as

$$\sigma = \frac{N_A P}{\sqrt{2\pi MRT}} \tau_0 e^{\Delta H_{ads}/RT} \tag{5.35}$$

From the knowledge of ΔH_{ads}, P, and T, σ can be estimated. We assume that the adsorbed gas atoms undergo complete thermal accommodation on the surface: $T = T_s = T_g$.

The gas atoms adsorbed on the surface at a given temperature, T and pressure P are either localized at well-defined surface sites or mobile (i.e., they possess translational degrees of freedom along the surface). The structure of the adsorbed layer can be studied by low-energy electron diffraction. Although the surface structure of adsorbed gases during physical adsorption should provide one with important information about the nature of the adsorbed layer, such studies have been carried out only recently for only a few systems.

5.4.2. THE SURFACE STRUCTURE OF WEAKLY ADSORBED GASES

Ordered $(\sqrt{3} \times \sqrt{3})$-R 30° surface structure of xenon on the (0001) graphite surface was observed at 90°K and at 10^{-3} torr by low-energy electron diffraction.[24] LEED studies have revealed the presence of an ordered argon (1×1) surface structure on the (110) crystal face of niobium at 20°K.[25] However, LEED studies of the adsorption of several gases (O_2, Ar, Xe, C_2H_4) on the (110) and (100) faces of silver single crystals at temperatures of 100–270°K indicated that the adsorbed gases are disordered.[26] It appears that the adsorbed gas may undergo order–disorder transformations as a function of temperature and pressure (surface coverage). Studies of the surface structure of weakly adsorbed gases by using LEED in the pressure range 10^{-9}–10^{-3} torr and at temperatures below 100°K are in progress in several laboratories, and it is hoped that a great deal of structural information will become available in the near future.

5.4.3. ADSORPTION ISOTHERMS

Most of our information about the nature of the weakly adsorbed gas layer comes from macroscopic studies of the amount of gas adsorbed on the surface σ (surface coverage), as a function of gas pressure P at a given temperature. The σ–P curves are called *adsorption isotherms*. The adsorption isotherms are used primarily to determine thermodynamic parameters that characterize the adsorbed layer (heats of adsorption and the entropy and heat capacity changes associated with the adsorption process) and to determine the surface area of the adsorbing solid. The latter measurement is of great technical importance because of the widespread use of porous solids of high surface area in various industrial processes. The effectiveness of participation by porous solids in a surface reaction is often proportional to the surface area of the solids. The simplest adsorption isotherm at a constant temperature is obtained from Eq. (5 35), which we can rewrite as

$$\sigma = kP \tag{5.36}$$

where $k = N_A(2\pi MRT)^{-1/2}\tau_0 \exp(\Delta H_{ads}/RT)$. Thus the coverage is proportional to the first power of the pressure at a given temperature if the model for adsorption that led to the formulation of Eq. (5.36) is correct. That is, the adsorbed gas atoms do not interact with each other and we have an unlimited number of surface sites at which adsorption can occur. It is also assumed that the adsorption energy ΔH_{ads} is the same for all of the molecules. Equation (5.36) is unlikely to be suitable to describe the overall adsorption process. Nevertheless, it approximates the adsorption isotherms for many real systems at low pressures ($<10^{-5}$ torr) and at high pressure (10 torr) at the initial stages of adsorption. The adsorption isotherms of argon on silica gel which obey Eq. (5.26) are shown in Fig. 5.12.

Langmuir has derived a different adsorption isotherm by assuming that adsorption is terminated upon completion of a monomolecular adsorbed

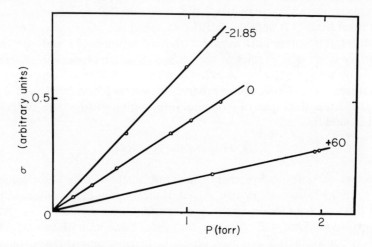

Figure 5.12. *Adsorption isotherms of argon on silica gel.*

gas layer.[27] This he has done by asserting that any gas molecule that strikes an adsorbed atom must reflect from the surface. All the other assumptions (i.e., homogeneous surface and noninteracting adsorbed species) used to obtain Eq. (5.36) were also maintained. If σ_0 is the surface coverage in the completely covered surface, the number of surface sites available for adsorption, after adsorbing σ molecules, is $\sigma_0 - \sigma$. Of the total flux F incident on the surface, a fraction $(\sigma/\sigma_0)F$ will strike molecules already adsorbed and therefore reflected. Thus a fraction $(1 - \sigma/\sigma_0)F$ of the total incident flux will be available for adsorption. Equation (5.34) should then be modified as

$$\sigma = \left(1 - \frac{\sigma}{\sigma_0}\right)F\tau \tag{5.37}$$

which can be rearranged to give

$$\sigma = \frac{\sigma_0 F\tau}{\sigma_0 + F\tau} = \frac{\sigma_0 kP}{\sigma_0 + kP} \tag{5.38}$$

By writing $\Theta = \sigma/\sigma_0$, where Θ is often called the "degree" of covering, Eq. (5.38) can be rewritten as

$$\Theta = \frac{k'P}{1 + k'P} \tag{5.39}$$

where $k' = k/\sigma_0$. It can be seen from Eq. (5.38) that at low pressures $k'P$ may be neglected in comparison with 1 in the denominator, and Eq.

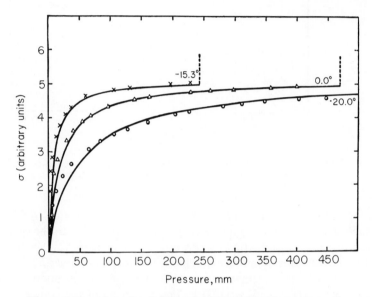

Figure 5.13. *Adsorption isotherms of ethyl chloride on charcoal.*

(5.36) is obtained. The adsorption isotherm of ethyl chloride on charcoal, which appears to obey an equation of the form of Eq. (5.38), is shown in Fig. 5.13. Equation (5.38) can be rearranged to give

$$\frac{1}{\sigma} = \frac{1}{kP} + \frac{1}{\sigma_0} \tag{5.40}$$

Therefore, a linear Langmuir plot is obtained by plotting $1/\sigma$ against $1/P$. Such a plot is shown for the adsorption of oxygen, carbon monoxide, and carbon dioxide on silica in Fig. 5.14.

There are several other derivations of the Langmuir adsorption isotherm from statistical mechanics and thermodynamics. Although the model is physically unrealistic for describing the adsorption of gases on real

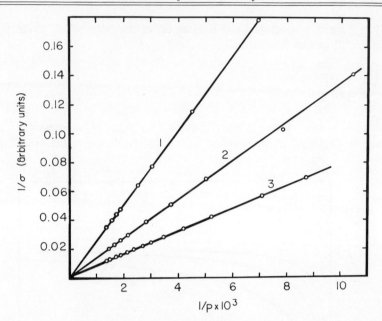

Figure 5.14. *Adsorption isotherm of (1) oxygen, (2) carbon monoxide, and (3) carbon dioxide on silica plotted as 1/σ versus 1/P.*

surfaces, its success, just like the success of other adsorption isotherms also based on different simple adsorption models, is due to the relative insensitivity of macroscopic adsorption measurements to the atomic details of the adsorption process   Thus the adsorption isotherm provides one with useful approximate values of the important adsorption parameters σ and ΔH_{ads} and permits the determination of the surface area. Another frequently used adsorption model that allows for adsorption in multilayers where gas atoms or molecules may adsorb on top of already adsorbed molecules was proposed by Brunauer, Emmett, and Teller[28] (BET). With the exception of the assumption that the adsorption process terminates at monolayer coverage, they have retained all other assumptions made in deriving the Langmuir adsorption isotherm. The BET model leads to a two-parameter adsorption equation of the form

$$\frac{P}{\sigma(P_0 - P)} = \frac{1}{\sigma_0 c} + \frac{c - 1}{\sigma_0 c} \frac{P}{P_0} \tag{5.41}$$

where P_0 is the saturation pressure of the vapor at which an "infinite" number of layers can be built up on the surface and c is a constant at a given temperature and is an exponential function of the heat of adsorption of the first layer and the heat of condensation or liquefaction of the vapor ΔH_L [$c \propto \exp (\Delta H_{\mathrm{ads}} - \Delta H_L)/RT$]. A plot of the quantity $P/\sigma(P_0 - P)$ versus P/P_0 should yield a straight line with slope $(c - 1)/\sigma c_0$ and intercept $1/\sigma_0 c$. The adsorption isotherm of nitrogen on titanium dioxide (anatase), which obeys Eq. (5.41), is shown in Fig. 5.15.

Another useful method of surface-area determination was suggested by Harkins and Jura.[29] It uses the linear relationship between the change

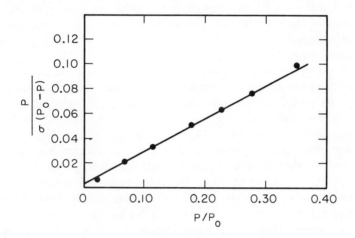

Figure 5.15. *Adsorption isotherm of nitrogen on titanium dioxide that obeys the BET equation.*

in surface tension of the clean solid upon adsorption of a gas or liquid and the surface area occupied by the adsorbed molecules [similar to the relationship between the surface work and the change in surface area as given by Eq. (2.9)]. This relationship may be transformed into an equation of the form

$$\log \frac{P}{P_0} = B - \frac{A}{\Theta} \tag{5.42}$$

Thus for any given system a plot of $\log P/P_0$ versus $1/\Theta$ should give a straight line of slope $-A$. The surface area is proportional to $A^{1/2}$. This method permits the determination of relative surface areas of adsorption for a given gas–solid system. There appears to be good agreement between the surface areas determined by the Harkins–Jura and the BET methods.

The adsorption isotherm yields the amount of gas adsorbed on the surface, i.e., the product of the average area occupied per molecule and the surface area. Unless the molecular area occupied by the adsorbed gas is known, the adsorption isotherm yields only relative surface areas rather than the absolute values. This is the reason for using only one gas (nitrogen or krypton) to determine the surface areas of different solids. However, Harkins and Jura[30] developed an absolute method to determine the surface area of nonporous solids by using an immersion calorimeter. If the finely divided crystal is coated with the adsorbed film (gas or liquid) and then is dropped into a liquid inside a calorimeter, the heat of immersion divided by the surface free energy of the liquid gives the surface area of the solid directly. By using this absolute method and the BET or other adsorption isotherms for the same system, the average area occupied by a molecule of a given gas can be obtained. For nitrogen this was found to be 16.2 Å^2 (and for krypton 25.6 Å^2) for a variety of surfaces and was adopted as a standard for surface-area determinations.[31]

There have been several theoretical models[32,33] proposed that attempt to give a more realistic description than the Langmuir and BET models of the gas–surface interactions that lead to physical adsorption. The variable parameters in these models are the interaction potential, the structure of the adsorbed layer (mobile or localized monolayer or multilayer), and the structure of the surface (homogeneous or heterogeneous, number of nearest neighbors). Unfortunately, many of these different detailed models of adsorption yield similar theoretical adsorption isotherms. Thus it is essential to obtain supplementary experimental information, in addition to the adsorption isotherms, before a more meaningful theory of physical adsorption can be developed. These experiments involve (1) the determination of the surface structure of weakly adsorbed gases on single crystal surfaces by LEED. Ordering in the adsorbed layer may also have an effect on surface-area determination, (2) studies of their order–disorder phase transitions by LEED at low temperatures, and (3) determination of adsorption isotherms and the heats of adsorption as a function of coverage for different gases on ordered single-crystal surfaces. So far, such studies have been carried out for only a few systems.

5.4.4. ADSORPTION ISOTHERMS ON SINGLE-CRYSTAL SURFACES

Adsorption isotherm studies, in general, are carried out on porous solids of high surface area ($\sim10^2$ m²/g) to optimize the accuracy of volumetric measurements. In order to uncover the role of the structure of the solid surface and of surface heterogeneities in the adsorption process, it is essential to measure adsorption isotherms on clean, well-defined, preferably single-crystal surfaces. Adsorption studies on freshly crushed or deposited crystal surfaces have been carried out by the use of microbalance.[34] Recently, ellipsometry techniques[26] were found to be useful in measuring adsorption isotherms on single-crystal surfaces. The amplitude and phase of polarized light reflected from a metal single crystal surface change upon adsorption of a gas in fractional monolayer quantities. The change in phase and the change in amplitude of the reflected polarized light appear to be a linear function of the "optical thickness," which is related to the average thickness of the adsorbed layer. By measuring the optical thickness as a function of gas pressure over the single crystal, the adsorption isotherm could be obtained. From the adsorption isotherms obtained at different surface temperatures, the heats of adsorption as a function of coverage could be calculated. Such data have been reported for the adsorption of Kr, Xe, O_2, CH_4, C_2H_2, and C_2H_4 on the (110) and (100) faces of silver crystals. Although contamination of the surface by unwanted condensates can be difficult to control, such measurements promise to yield detailed information about the adsorption process and should aid in the development of realistic theoretical models of physical adsorption.

Another method to measure adsorption isotherms and the heat of adsorption of gases on single crystal surfaces correlates the change of work function with the gas pressure over the surface and the surface temperature.[35] The work-function change (increase) was found to be linearly proportional to the surface coverage of carbon monoxide on the Pd(100) crystal face. The change of work function of the palladium surface was monitored as a function of temperature at a constant CO pressure. Then the measurement was repeated at several gas pressures. From the data a family of curves on a $\ln P_{CO}$ versus $1/T^\circ K$ plot could be obtained for various values of the work-function change. The heat of desorption of CO at constant coverage, ΔH_{des}, on the palladium surface is then obtained from the Clausius–Clapeyron equation,

$$[d(\ln P)/d(1/T)]_{\theta=\text{const}} = -\frac{\Delta H_{des}}{R}.$$

The heat of desorption was found to decrease markedly with increasing coverage. The assumption that the change of work function is a unique function of coverage and independent of temperature is borne out by the experimental results.

5.4.5. INTEGRAL AND DIFFERENTIAL HEATS OF ADSORPTION AND OTHER THERMODYNAMIC PARAMETERS OF THE ADSORBED LAYER

When a single atom or molecule is adsorbed on a clean surface, heat is liberated; i.e., the adsorption process is always exothermic. The total or integral heat of adsorption ΔH_{ads}, measured for the adsorption of N molecules in the overall adsorption process, is proportional to the number of adsorbed molecules N and to the heat of adsorption per molecule, q_{ads}:

$$\Delta H_{ads} = N q_{ads} \tag{5.43}$$

Frequently one is interested in measuring the differential heat of adsorption $(d\,\Delta H_{ads}/dN)_T$, which is the increase in the heat liberated by adsorbing an additional amount of gas, dN. If q_{ads} is independent of the amount of gas already adsorbed on the surface, a condition that is assumed in deriving the isotherms given by Eqs. (5.36), (5.39), and (5.41), the differential heat of adsorption is independent of coverage and remains constant as a function of coverage. In reality, the heat of adsorption changes with the amount of gas adsorbed on the surface due to the heterogeneity of the surface. Thus it is useful to differentiate ΔH_{ads} with respect to N to obtain

$$\Delta H_{ads}^{diff} = \left(\frac{d\,\Delta H_{ads}}{dN} \right)_T = q_{ads} + N \left(\frac{dq_{ads}}{dN} \right)_T \tag{5.44}$$

If the differential heat of adsorption ΔH_{ads}^{diff} is measured under isothermal conditions it is commonly called *isothermal* or *isosteric* heat of adsorption.[23]

The integral and differential heats of adsorption are determined by measuring the adsorption isotherms for a given system at different temperatures. From the data, the equilibrium pressures necessary to obtain the *same* coverages at the different temperatures are determined. From the slope of the plots of $\log P(\text{constant } \Theta)$ versus $1/T$, the differential isosteric heats of adsorption for a given coverage are determined. Variation of the differential isosteric heat of adsorption ΔH_{ads}^{diff}, with degree of coverage Θ, for several systems is shown in Fig. 5.16. The ΔH_{ads}^{diff} versus Θ curves may be extrapolated to zero coverage to obtain q_{ads}, which is a direct measure of the interaction energy between the gas molecule and the surface. As the coverage is increased the heat of adsorption may increase or decrease because of (1) interactions between the adsorbed molecules and (2) the availability of different surface sites for adsorption on the heterogeneous surface. At high coverages ($\Theta > 1$) the heat of adsorption approaches the heat of condensation (liquefaction) of the adsorbed gas indicating that the surface forces no longer play an important role in determining the properties of the adsorbed layer. It should be noted that because of surface forces, at low coverages the heat of adsorption is several times the value of the heat of liquefaction.

The isosteric heat of adsorption $\Delta H_{\text{ads}}^{\text{diff}}$ for a given coverage changes with temperature if the heat capacity of the adsorbed gas on the surface, C_S, is different from that in the gas phase, C_P (which is always the case since there is a loss of at least one translational degree of freedom for the adsorption of a mobile layer, or a loss of three translational degrees of freedom for the adsorption of a localized monolayer). The heat capacity

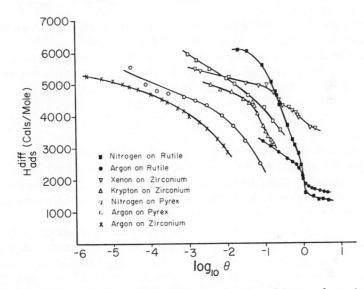

Figure 5.16. *Isosteric heats of adsorption as a function of degree of covering for several systems.*

of the adsorbed gas is given by

$$C_S = C_P + \frac{\partial(\Delta H_{\text{ads}}^{\text{diff}}/n_s)}{\partial T} \tag{5.45}$$

where n_s is the number of moles of adsorbed gas. The entropy of the adsorbed gas, S_S, can similarly be expressed as

$$S_S = S_{T,P_0} - \frac{\Delta H_{\text{ads}}/n_s}{T} - \frac{R}{n_s}\int_0^{n_s} \ln\frac{P}{P_0}\,dn_s \tag{5.46}$$

where S_{T,P_0} is the entropy of the gas at temperature T and pressure P_0, the second term on the right side of Eq. (5.46) is the entropy change in desorbing the gas completely at equilibrium, and the third term is the entropy change in bringing the desorbed gas from the equilibrium pressure P to a reference pressure P_0.

In studies of the thermodynamics of adsorbed layers, it is useful for certain systems (e.g., in studies of monomolecular surface films on the liquid surfaces) to define the surface pressure Π as

$$\Pi = \gamma_0 - \gamma \tag{5.47}$$

where γ_0 is the surface tension of the clean surface and γ is its surface tension in the presence of the adsorbed layer. For insoluble layers, which are dilute and mobile so that they approximate the properties of a two-dimensional surface gas (gaseous film), we can write the two-dimensional equivalent of the ideal gas law

$$\Pi\mathscr{A} = n_s RT \tag{5.48}$$

For example, certain proteins (hemoglobin, ovalbumin) on water seems to obey Eq. (5.48). As the concentration of the surface layer increases, Eq. (5.48) has to be modified just like the ideal gas law at high pressures. The reader interested in the thermodynamics of adsorbed monolayers is referred to detailed treatments of the subject.[35,36]

The surface pressure due to adsorbed films on a liquid surface may be measured by "film balances" of different types, which measure changes of surface tension due to spreading or compression of the surface films on liquids.[37,38] Such measurements are of great importance in testing the properties of soap and detergent solutions. Their role as cleaning agents requires that they alter the surface tension of water to facilitate the dissolution of solids of different kinds. Insoluble monomolecular films of organic acids spread evenly on the surfaces of lakes and reservoirs are often used to reduce the evaporation rate of water by as much as 40 per cent by making vaporization diffusion limited through the surface film. In this way water may be conserved and used for irrigation of arid areas.

5.5. Strong Interactions of Gas Atoms with Surfaces

5.5.1. INTRODUCTION

When gas atoms or molecules that experience a strong attractive potential ($\Delta H_{\mathrm{ads}} \gg RT_s$) strike a surface they are likely to stick to the surface for times much longer than lattice vibration times of surface atoms. One can calculate, by using Eq. (5.2), that for $\Delta H_{ads} \sim 15$ and 20 kcal/mole the residence times are on the order of 10^{-2} and 10^2 sec, respectively, at $300°K$ assuming that $\tau_0 = 10^{-12}$ sec. In these circumstances the energy-transfer probability between the incident gas and the surface is greatly increased. Complete thermal accommodation of the incident molecular

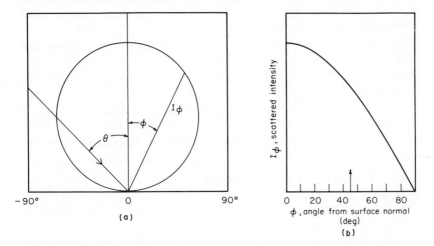

Figure 5.17. *Cosine distribution of scattered atoms displayed in (a) polar and (b) rectilinear coordinate systems.*

beam can often occur and then the molecules are reemitted from the surface with angular and energy distributions corresponding to the surface temperature T_s. The flux of molecules F (molecules/cm² sec) that emerges from the surface after complete energy accommodation is given from the kinetic theory of gases as $F = \frac{1}{4}n\bar{c}$, where $\bar{c}$ is the average velocity corresponding to the surface temperature and n is the number of molecules per unit volume. The flux of molecules that emerges from the surface in a solid angle $d\omega$ at a scattering angle ϕ relative to a surface normal is[39]

$$dF = \frac{d\omega}{4\pi} n\bar{c} \cos \phi \qquad (5.49)$$

which is the well-known cosine distribution from the kinetic theory. Thus the flux decreases as $\cos \phi$ away from the surface normal. A typical cosine distribution of emitted atoms plotted in both polar and rectilinear coordinate systems is shown in Fig. 5.17.

In most molecular-beam-scattering experiments, however, there is only partial energy transfer between the surface and the gas molecules. The incident molecules are rescattered from the surface before they can be fully accommodated. Nevertheless, the direct energy transfer between the surface phonons and some of the internal states of the molecules, rotational or vibrational states, for example, changes the energy distribution and the angular distribution of the scattered beam markedly as compared with their distribution after impulsive collision during weak interaction with the surface.

The incident atom or molecule may also undergo strong interaction with the surface, which leads to a chemical reaction. Reactive scattering may be divided into two classes: (1) those that lead to chemical reactions between the gas and the surface in which the surface acts as one of the reactants (e.g., the reaction between an oxygen molecular beam and a metal surface where the rescattered products are metal oxide molecules), and (2) surface chemical reactions in which the incident gas molecule undergoes dissociation (e.g., $N_2O \xrightarrow[surf]{w} N_2 + O$) or other chemical rearrangements (association, hydrogenation, etc.), but the surface may participate only as a "catalyst" by providing reaction sites, mobile charge carriers, etc., while remaining unchanged structurally and chemically during the reaction.

It should be possible to deduce the energy-transfer processes that take place during the strong surface interactions (and reactive scattering) of molecules by monitoring the energy and angular distributions and composition of the scattered beams. So far only a few investigations of this type have been reported; therefore, we will give only examples of the molecular-beam-scattering studies of these different types of strong interactions. It is hoped that reactive-scattering studies will be carried out in the near future since the understanding of the nature of energy transfer in surface reactions is a problem of great importance in surface science.

5.5.2. SCATTERING OF DIATOMIC MOLECULES FROM SURFACES

For diatomic molecules, energy transfer between surface phonons and the rotational and vibrational energy states is also possible in addition to direct surface phonon-translational energy transfer. That such energy transfer takes place efficiently is shown by studies of surface scattering of H_2, D_2, and HD molecular beams.[40] The angular distribution of these molecules scattered by silver (111) surfaces is shown in Fig. 5.18. H_2 is scattered strongly in the specular angle ($\phi = \theta$), indicating that it does not participate in energy transfer with the surface to any great extent. On the other hand, D_2 and HD beams have only a weak specular component after scattering, which signifies marked energy transfer during the surface collision. It appears that D_2 and HD undergo rotational excitation because of their low-lying rotational energy states. For these molecules the allowed rotational transitions require energies in the

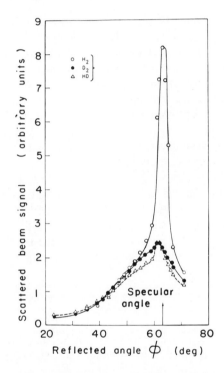

Figure 5.18. *Angular distribution of* H_2, D_2, *and* HD *molecular beams from an oriented silver (111) film.*

range 250–520 cal/mole, which is in the range of thermal energies of vibrating surface atoms at the experimental temperature $T_s = 570°K$. For H_2 molecules, however, allowed rotational transitions require energies in excess of 10^3 cal/mole, which can be transferred only by less probable multiphonon excitations at these surface temperatures. This interpretation of the scattering behavior of these molecules is supported by the observation that at high beam temperature ($T_g \approx 1500°K$), D_2 scatters specularly similar to H_2 beams. At these beam temperatures the incident molecules already have a large population in the higher rotational states; thus the energy-transfer probability between the surface phonons and the molecular rotational states is decreased.

5.5.3. REACTIVE SCATTERING OF ATOMIC AND MOLECULAR OXYGEN BEAMS FROM (100) AND (111) CRYSTAL FACES OF SILICON AND GERMANIUM

The interaction of beams of oxygen atoms and oxygen molecules with silicon and germanium surfaces yields volatile monoxides, SiO and GeO. The reaction probabilities were calculated from the observed weight-loss rates of the crystals for known fluxes of the incident oxygen beams.[41] The reaction probability of the oxygen atoms with both solids was in the range 0.2–0.3 for surface temperatures between 830 and 1110°K. The reaction probability for oxygen molecules was much lower—0.02–0.03 for both semiconductors in the same temperature range. Electronically excited species were absent in the incident beams. The reactions required little activation energy, as indicated by the virtual independence of the reaction probabilities from surface temperature. The difference in reactivity appears to be due to the requirement that both oxygen atoms in the oxygen molecule interact simultaneously with the germanium or silicon surface atoms; i.e., the interaction probability depends on the orientation of the oxygen molecule as it approaches the surface. For oxygen atoms there is no such orientational criterion.

5.5.4. DISSOCIATION OF HYDROGEN MOLECULAR BEAMS ON TANTALUM

The dissociation of hydrogen molecules on hot tantalum surfaces was studied at surface temperatures in the range 1100–2600°K.[42] Dissociation products (H atoms) are not detectable at low temperatures, whereas they constitute about 30 mole per cent of the incident molecular beam at the highest temperatures. The scattered beam was monitored at a fixed angle, so variation of the angular distribution of the scattered molecules as a function of surface temperature or other variables has not been studied. The unreacted H_2 molecules were scattered in phase with the incident beam, which indicates incomplete energy accommodation on the surface and relatively short residence times ($\ll 10^{-3}$ sec). The hydrogen atoms produced by the dissociation reaction appeared to undergo complete thermal accommodation on the hot tantalum surface.

By using high-beam-temperature molecular hydrogen ($T_g = 1150$–1800°K), the efficiency of producing hydrogen atoms in the surface collision process was *increased* as compared with that for a beam at $T_g = 300°K$. The dependence of the atomization surface reaction on beam temperature indicated that it is an activated process with an apparent activation energy of 1.4 kcal/mole. This is compared with the dissociation energy of H_2 in the gas phase, which is about 103 kcal/mole. The use of beams of D_2 and HD mixed with H_2 beams revealed that no isotope exchange takes place. Thus the atoms, once formed, do not recombine into molecules but desorb as atoms from the tantalum surface. The desorption rate of hydrogen atoms appears to be the rate-limiting step in the reaction for obtaining atomic hydrogen since the activation energy for H-atom desorption is $\sim$75 kcal/mole. In contrast to their low desorption rate, the surface diffusion of hydrogen atoms on tantalum appears to be rapid and may not require activation energy.

5.5.5. DISSOCIATION OF N$_2$O BEAMS ON CLEAN AND CARBON COVERED PLATINUM

The dissociation of molecular beams of N_2O on the (100) crystal plane of platinum and on a carbon covered platinum surface has been studied in the temperature range $T_s = 900$ to $1500°K$ in an ultra high vacuum chamber.[43] The angular distribution of the scattered beams was monitored with a quadrupole mass spectrometer, the target surface structure by low energy electron diffraction, and the surface composition by Auger electron spectroscopy. The effect of surface temperature and surface structure on the surface decomposition of N_2O has been studied.

N_2O may undergo a variety of different chemical surface reactions upon scattering from platinum surfaces. A clean platinum surface seems to dissociate N_2O only poorly; at the surface temperatures employed in this work only a few per cent of the incident molecules undergo bond breaking. This reaction appears to be endothermic and shows only a very small temperature dependence. The incident molecules that dissociate (to yield N_2, NO and O) are fully accommodated on the surface before reemission as indicated by the cosine angular distribution of the scattered beam. On the carbon covered platinum (100) surface, the scattering process appears to be entirely different. Due to the interaction between N_2O and surface carbon, the surface reaction can be strongly exothermic (to yield CO, CO_2 or CN). As a result, evidence for direct reactive scattering has been detected. The NO molecules are scattered without energy accommodation between the incident beam and the surface as indicated by the non-cosine angular distribution of the scattered beam. Direct scattering is commonly observed in studies of chemical reactions between crossed molecular beams that are exothermic and exoergic.

5.6. *Chemisorption*

5.6.1. INTRODUCTION

We have seen that the strong attractive interaction potential between the adsorbed gas and surface atoms may give rise to residence times $>10^{-2}$ sec for heats of adsorption ΔH_{ads} greater than 15 kcal/mole. Thus, even for small gas fluxes ($\leqslant 10^{-6}$ torr), large surface coverages can be obtained at 300°K or at higher temperature [according to Eq. (5.34)]. The adsorption of gases that undergo strong interaction with the surface ($\Delta H_{ads} > 15$ kcal/mole) is frequently called *chemisorption*. An important part of most studies of gas–surface interactions and many chemical surface reactions is to investigate the structural, thermodynamic, and electrical properties of the chemisorbed layer. These fundamental properties provide us with information about the nature of the surface chemical bond between the adsorbed gas molecules and atoms in the solid surface. Since the surface reaction of gases is likely to take place in the adsorbed state, investigation of the chemisorbed layer leads to a better understanding of chemical surface reactions.

We will first discuss the surface structure of chemisorbed gases. Then the adsorption isotherms and the heats of chemisorption of different chemisorbed gases will be reviewed. Finally, the electrical properties of adsorbed gases on metal and semiconductor surfaces, which are studied primarily by work-function and surface conductance measurements, will be discussed.

5.6.2. THE SURFACE STRUCTURE OF CHEMISORBED GASES

The structure of chemisorbed gases has been studied extensively by low-energy electron diffraction. LEED studies have shown that chemisorption predominantly yields ordered structures on single-crystal surfaces (commonly called *substrates*). The surface structures that form depend, to a great extent, on the symmetry of the substrate; the chemistry and size of the adsorbed gas molecule; and, in some cases, the surface concentration of the adsorbate, which may be controlled by the gas partial pressure over the surface.

In Tables 5.4 through 5.7 the surface structures of chemisorbed gases on different crystal surfaces are listed for those solids that have been studied by low-energy electron diffraction.

Most of the experimental data were accumulated on the surface structures of adsorbed gases that form on the lowest index (highest density) faces of face-centered cubic and body-centered cubic crystal surfaces

Table 5.4. *Surface Structures of Chemisorbed Gases on Face-centered Cubic Crystal Surfaces*

Surface	Absorbed gas	Surface structure	Reference
Ni(111)	O_2	(2×2)-O	60, 70, 71
		$(\sqrt{3} \times \sqrt{3})$-R 30°-O	69, 72
	CO	(2×2)-CO	70
	Dissociates to $C + CO_2$	$(16\sqrt{3} \times 16\sqrt{3})$-R 30°-C +	72
		$(2 \times \sqrt{3})$-CO_2	72
	H_2	(1×1)-H	70
		Disordered	73
	N_2	Not adsorbed	74
	CO_2	(2×2)-CO_2	72
		$(2 \times \sqrt{3})$-CO_2	72
	Dissociates in electron beam to	$(16\sqrt{3} \times 16\sqrt{3})$-C +	72
		$(\sqrt{3} \times \sqrt{3})$-R 30°-O	72
	C_2H_4	(2×2)-C_2H_4	75
	C_2H_6	(2×2)-C_2H_6	75
	C_3H_6	(2×1)-C_3H_6	75
		$(\sqrt{7} \times \sqrt{7})$R 19°-C	75
			69, 77, 78
Ni(100)	O_2	(2×2)-O	79, 80, 81, 82,
		$c(2 \times 2)$-O	69, 83, 84, 85
			82, 83, 86
	CO	$c(2 \times 2)$-CO	73
	H_2	Disordered	73
	N_2	Not adsorbed	
Ni(110)	O_2	(2×1)-O	70, 69, 85, 88, 89
		(3×1)-O	69, 73, 78, 90
		(5×2)-O	69
		(5×1)-O	69
	CO	(1×1)-CO	69, 91
	H_2, D_2	(1×2)-H	73, 91, 92, 93
	H_2O	(2×1)-H_2O	92
	C	$c(2 \times 2)$-C	92
Ni(210)	I_2	Facet to Ni(540)	94
Pt(111)	O_2	(2×2)-O	95, 96, 97, 98
	$H_2 + O_2$	$(\sqrt{3} \times \sqrt{3})$-R 30°	71
	CO	$c(4 \times 2)$-CO	99
	C_2H_2	(2×1)-C_2H_2	99
	C_2H_4	(2×1)-C_2H_4	99

Table 5.4. *Surface Structures of Chemisorbed Gases on Face-centered Cubic Crystal Surfaces* (*Continued*)

Surface	Adsorbed gas	Surface structure	Reference
	C_3H_6	(2×1)-C_3H_6	99
(*cis* and *trans*)C_4H_8 (2-butene)		(2×2)-C_4H_8)	99
C_4H_6 (butadiene)		(2×2)-C_4H_6	99
C_4H_8 (isobutylene)		$(\sqrt{7} \times \sqrt{3})$-R $13.9°$	99
Pt(100)-(5 × 1)	O_2	(1×1)-O	96
	H_2	(2×2)-H	100, 101
	CO	c(4×2)	98, 99, 100, 102
		$(3 \times 2 \sqrt{2})$-R $245°$	98, 99, 100, 102
		$(\sqrt{2} \times \sqrt{5})$-R $45°$	100, 101
	C_2H_2	c(2×2)-C_2H_2	99, 100
	C_2H_4	c(2×2)-C_2H_4	99, 100
	C_3H_6	Disordered	99
(*cis* and *trans*)C_4H_8 (2-butene)		Disordered	99
C_4H_8 (isobutylene)		Disordered	99
C_4H_6 (butediene)		Disordered	99
	CO + H_2	c(2×2)-(CO + H_2)	100, 101
Pt(110)	O_2	(1×2)	97
		(2×4)	97
Pd	CO	Disordered	103
		c(4×2)-CO	103
		compressed	103
Cu(111)	O_2	(11×5)-R $5°$-O	107
Cu(100)	O_2	c(2×2)-O	105
		(1×1)-O	105
		(2×1)-O	105
	N_2	(1×1)-N	106
Cu(110)	O_2	(2×1)-O	105, 107
		c(6×2)-O	107
Cu(035)	O_2	(1×1)-O	105
Cu(014)	O_2	(1×1)-O	105
Al(100)	O_2	Disordered	109, 110, 111
Rh(100)	O_2	(2×8)-O	112
	CO	(4×1)-CO	112
Rh(110)	O_2	c(2×4)-O	113, 114
		c(2×6)-O	113, 114
		c(2×8)-O	113, 114
		(2×2)-O	113, 114
		(2×3)-O	113, 114
UO_2(111)	O_2	(3×3)-O	115
		$(2\sqrt{3} \times 2\sqrt{3})$-R $30°$-O	115

Table 5.5. *Surface Structures of Chemisorbed Gases on Body-centered Cubic Crystal Surfaces*

Surface	Adsorbed gas	Surface structure	Reference
W(110)	O_2	(2×1)-O	85, 116, 117
		$c(14 \times 7)$-O	85, 116, 117
		$c(21 \times 7)$-O	117
		$c(48 \times 16)$-O	117
		$c(2 \times 2)$-O	117
		(2×2)-O	117
		(1×1)-O	117
	$O_2 + CO$ coadsorption	$c(11 + 5)$-CO + C_2	118
	CO	Disordered	119
		$c(9 \times 5)$-CO	
	CH_4	(15×3)-C	120
		(15×12)-C	
W(111)	O_2	To (211) facets	121
	CH_4	(6×3)-C	120
W(211)	O_2	(2×1)	121
		(4×3)	121
		(1×2)	121
		$(1 \times n)$-O, $n = 1, 2, 3, 4$	122, 123
	NH_3 thermal breakup	$c(4 \times 2)$-NH_2, 12% stretch	124
	CO	$c(6 \times 4)$-CO	125
		(2×1)-CO	125
		$c(4 \times 2)$-CO	125
W(100)	O_2	(4×1)-O	126
		(2×1)-O	126
	CO	$c(2 \times 2)$-CO	126, 127
	N_2	$c(2 \times 2)$-N	126
	CH_4	(5×1)-C	120
	NH_3	Disordered	121
		$c(2 \times 2)$-NH_2	121
		(1×1)-NH_2	121
	$(CO + N_2)$	(4×1)-$(CO + N_2)$	127
	H_2	$c(2 \times 2)$-H	128, 129
		(2×5)-H	129
		(4×1)-H	129
Ta(110)	O_2	(3×1)-O	130
		(3×2)-R $18°16'$-O	131
		Oxides	131
	N_2	Not Adsorbed	131
	H_2	(1×1)-H	131

Table 5.5. *Surface Structures of Chemisorbed Gases on Body-centered Cubic Crystal Surfaces (Continued)*

Surface	Adsorbed gas	Surface structure	Reference
	CO	Disordered	131
		Decomposition to	
		$C + CO_2$	131
Ta(112)	O_2	(3×1)-O	131
		Oxides	131
	N_2	Nitride form epitaxially	131
		on (113) planes	
	H_2	(1×1)-H	131
	CO	Disordered	131
		Decomp. to $C + CO$	131
Nb(110)	O_2	(3×1)-O	132
		(3×2)-R $18°16'$-O	132
	H_2	(1×1)-H	133
V(100)	O_2	(1×1)-O	134
		(2×2)-O	134
	H_2	Disordered	134
V(110)	CO	Disordered	132
Cr(100)	O_2	(2×2)-O	93
	CO	(2×2)-CO	93
	N_2	(2×2)-N	93
α-Fe(110)	O_2	$c(2 \times 2)$-O	135, 136
		$c(3 \times 1)$-O	135, 136
		$c(1 \times 5)$-O	135
		(2×8)-O	135
		FeO(111) (cubic)	136
		γ-Fe_2O_3 (spinel)	136
Mo(110)	O_2	(2×2)-O	137, 138, 139
		(2×1)-O	137, 138, 139
		(1×1)-O	138, 139
		$c(2 \times 2)$-O	139
	CO	(1×1)-CO	137, 139
		$c(2 \times 2)$-CO	91
	H_2	Adsorbed (no structure	137
		given)	
	CO_2	Disordered	91
Mo(100)	H_2	$c(4 \times 2)$-H	140
		(1×1)-H	140
	O_2	Disordered	141
		$c(2 \times 2)$-O	138, 139, 141
		(1×1)-O	141
		$\sqrt{5}(1 \times 1)$-R $\pm 26°34'$-O	139, 141
		(2×2)-O	141
	N_2	(1×1)-N	139
	CO	(1×1)-CO	139

Table 5.6. *Surface Structures of Chemisorbed Gases on Diamond Crystal Surfaces*

Surface	Adsorbed gas	Surface structure	References
Si(111)-(1 × 1)	O_2	(1 × 1)	142, 143
-(7 × 7)		Disordered	142, 143
	H_2S	(2 × 2)-S	144
	H_2	Not adsorbed	144
	H_2Se	(2 × 2)-Se	144
	I_2	(1 × 1)	144
	NH_3	(8 × 8)-N	145
	PH_3	$(6\sqrt{3} \times 6\sqrt{3})$-P	146
		(1 × 1)-P	146
		$(2\sqrt{3} \times 2\sqrt{3})$-P	146
Si(100)	O_2	(1 × 1)	142, 143, 147, 148
		(111) facets	142, 143, 147, 148
	I_2	(3 × 3)	149
C(diamond)(111)	O_2	Ordered?	150
	CO_2	?	150
C(graphite)(0001)	O_2	Not adsorbed	151
	CO	Not adsorbed	151
	H_2O	Not adsorbed	151
	I_2	Not adsorbed	151
	Br_2	Not adsorbed	151
C(diamond)(100)	O_2	Disordered	150
		Ordered	150
Ge(111)	O_2	(1 × 1)	142, 148
		Disordered	152
	I_2	(1 × 1)	152
Ge(100)	O_2	(1 × 1)	142, 148
		Disordered	142, 148
	I_2	(3 × 3)	152
Ge(110)	O_2	(1 × 1)	142, 148
		Disordered	142, 148

(although several experimental studies have been carried out on low-index diamond and close-packed hexagonal crystal surfaces). Most of the adsorbed gases studied so far have molecular dimensions smaller than or similar in size to the largest interrow distances in the substrate crystal surfaces. Thus these molecules or atoms may easily "fit" onto the surface without the need for overlapping several substrate atoms. The gases are oxygen, carbon monoxide, hydrogen, nitrogen, ammonia, and carbon dioxide; they may adsorb as molecules or in a dissociated form. Recently, however, longer chain olefins and aromatic hydrocarbons with

Table 5.7. *Surface Structures of Chemisorbed Gases on Hexagonal Close-packed Crystal Surfaces*

Surface	Adsorbed gas	Surface structure	References
Ti(0001)	O_2	(1×1)	148
	CO	(1×1)	148
Re(0001)	O_2	(2×2)-O(CO)	108
		(1×1)-O(CO)	108
	CO	(2×2)-CO	108
Be(0001)	O_2	Disordered	104
	CO	Disordered	104
	N_2	Not adsorbed	104
	H_2	Not adsorbed	104
CdS(0001)	O_2	Disordered	76

large molecular dimensions have also been studied, and their surface-structure-forming characteristics permit the detection of steric effects in surface ordering.

The simple abbreviated nomenclature used to identify the surface structure of adsorbed gases in Tables 5.4 to 5.7 has already been discussed in Chapter 1. For example, if the surface structure that forms in the presence of an adsorbed gas is characterized by unit-cell dimensions twice as large as the substrate unit cell it is called a (2×2) surface structure; i.e., the surface structure of butadiene (C_4H_6) on the (111) face of platinum is denoted Pt(111)-(2×2)-C_4H_6. All the surface structures listed were judged reproducible by the investigators.

The experimental data accumulated so far indicate several regularities or trends that are operative in the formation of surface structures of adsorbed gases on high-density crystal planes. It is possible to propose a set of rules that appear to govern the formation of ordered surface structures. It is hoped that the judicious application of these rules to other substrate–adsorbate systems that have not been studied will allow one to predict the surface structure that should form. These rules, although they are empirical and are formulated from the correlation of existing surface structural data, are nevertheless based on a strong physical chemical foundation. It appears that chemisorption leads to the formation of surface structures that exhibit maximum adsorbate–adsorbate and adsorbate–substrate interactions.

It should be noted that there are systems that represent exceptions to the rules of ordering proposed below. These exceptions will also be pointed out and discussed whenever possible.

5.6.3. RULES OF ORDERING OF CHEMISORBED GASES

RULE OF CLOSE PACKING: *Adsorbed atoms or molecules tend to form surface structures characterized by the smallest unit cell permitted by the molecular dimensions and adsorbate–adsorbate and adsorbate–substrate interactions. They prefer close-packing arrangements.* Inspection of Tables 5.4 to 5.7 reveals the absence of surface structures with large unit cells. The most dominant structures are those in which the unit-cell size is the same (1×1), or approximately twice as large as the substrate unit cell $[(2 \times 2)$, $c(2 \times 2)$, $(\sqrt{3} \times \sqrt{3})$, and $(2 \times 1)]$. The adsorbed atoms seem to pack as closely as allowed by the interactions between adsorbate molecules and the interaction with the substrate. The preferred close-packing arrangement indicates that the adsorbate–adsorbate interactions for the molecules, which were investigated by LEED, are just as strong as the adsorbate–substrate interaction and plays an important role in the formation of the surface chemical bond.

Adsorbed molecules whose dimensions are larger than the largest interrow spacing in the substrate surface also form the smallest possible unit cells. For example, propylene, 2-butenes, (*cis* and *trans*), iso-butylene, and butadiene form (2×1) and (2×2) surface structures on the (111) face of platinum. These are the smallest unit cells that can be formed compatible with size of the molecules. On the (100) surface of platinum, these olefins adsorbed in a disordered manner even though the surface concentrations are similar to those on the (111) face, as indicated by work-function measurements. It appears that if ordering under the close-packing conditions is difficult because of steric hindrances to rotation in the surface layer, the adsorbed layers will be disordered rather than forming an ordered surface structure with a large unit cell. This observation indicates that the adsorbed molecules listed in Tables 5.4 to 5.7 prefer close-packing arrangements on the surface of different substrates.

Certain difficulties arise when one attempts to apply this rule of ordering to all chemisorbed structures. There are molecules that exhibit more than one binding state on a given surface. Carbon monoxide, for example, has at least two binding states on the (100) surfaces of different face-centered cubic metals, as indicated by the variety of surface structures formed on that face as a function of temperature and partial pressure of carbon monoxide. On the other hand, on the (111) faces, one binding state seems to be preferred. Both the dissociated ammonia on the tungsten surface and carbon monoxide on the palladium (100) surface form structures that are stretched or compressed without changing the configuration of the adsorbed atoms as a function of temperature. The dissociation of the adsorbate may lead to changes in the structure and chemistry of the substrate that modify the surface structure. For example, dissociation of olefins or the disproportionation of CO to C and CO_2 leads to carbon deposition, which greatly affects the structure of the substrate and the adsorbate. Formation of nitrides or carbides at the surface would influence the nature of the adsorbate structure. These difficulties should be considered and taken into account when applying the close-packing rule to predict the nature of ordered surface structures.

In some experiments the diffraction pattern indicates the presence of large unit cells. For example, during the carburization of tungsten, both the (111) and (100) crystal faces exhibited a (5×1) surface structure. Surface-phase transformations can also yield (5×1) surface structures on gold, platinum, and iridium surfaces. A (7×7) surface structure is detected on the clean (111) surface of silicon, and large unit cells are detectable on germanium surfaces as well. Aluminum oxide, Al_2O_3, also exhibits a large surface unit cell. The appearance of diffraction patterns, which indicates large surface unit cells, is indicative of large mismatch between the surface layer and the underlying substrate (which produces coincidence lattices[44]; see Chapter 1). Surface-phase transformations or changes of valency of ions in the surface layer can give rise to large apparent unit cells frequently rotated by small angles with respect to the substrate unit cell (see Chapter 1).

RULE OF ROTATIONAL SYMMETRY: *Adsorbed atoms or molecules form ordered structures that have the same rotational symmetry as the substrate face.* The substrates, on which most of the adsorption studies have been carried out so far, show three different rotational symmetries. Some of the substrates exhibit sixfold rotational symmetry [(f.c.c. (111), b c.c. (111), diamond (111), hexagonal (0001)], others have a fourfold rotational axis [f.c.c. (100), b.c.c. (100)], and many have twofold rotational symmetries [f.c.c. (110), b.c.c. (110) and (211)]. Figure 1.19(c) shows the surface structures most frequently encountered for substrates that exhibit sixfold rotational symmetry. These are the (2×2), $[(\sqrt{3} \times \sqrt{3})R\text{-}30°]$, and the (1×1) surface structures. In every case the surface structures follow the rotational symmetry of the substrate. Figure 1.19(b) shows two surface structures $[(2 \times 2)$ and $c(2 \times 2)]$, which are observed most frequently on the substrates with fourfold rotational symmetry. It appears from inspecting Tables 5.4 to 5.7 that the $c(2 \times 2)$ surface structure is somewhat more prominent, as it leads to surfaces with higher coverage. Figure 1.19(a) shows the surface structures which were detected on crystal substrates showing twofold rotational symmetry. In addition to the $c(2 \times 2)$ structure, these are all $(n \times m)$, where $n \neq m$ surface structures. Their appearance reflects the fact that the magnitude of the unit-cell vectors in the x and y directions along the surface are either different [e.g., f.c.c. (110) and b.c c. (211) crystal faces] or if they are the same, subtend an angle of about 70° [b.c.c. (110)].

RULE OF SIMILAR UNIT CELL VECTORS: *Adsorbed atoms or molecules in monolayer thickness tend to form ordered surface structures characterized by unit cell vectors closely related to the substrate unit cell vectors. Thus the surface structure bears a greater similarity to the substrate structure than to the structure of the bulk condensate.* Only after the deposition of several atomic layers will the deposited structure adopt the surface structure of the pure condensed solid. Closer inspection of Tables 5.4 to 5.7 indicates that virtually all the surface structures of chemisorbed gases listed could be identified in terms of the unit-cell vectors of the substrate. All the surface unit cells of adsorbed gases can be given as some integral multiple of the substrate unit cell [i.e., (2 × 2) surface structure], or the rotated substrate unit cell [i.e., c(2 × 2) or ($\sqrt{3} \times \sqrt{3}$)-R 30°]. Thus it appears that the adsorbed molecules tend to prefer arrangements in which they are accommodated in the already existing surface structure by adopting the same periodicity that characterizes the substrate structure. This rule, although less important where chemisorption terminates at monolayer coverage, should have notable consequences if applicable to studies of multilayer chemisorption or epitaxy. Gradual distortion and mismatch should be observable in atomic layers of condensates deposited *on top* of the first condensate layer (which will have substrate-like structure) until lattice dimensions characteristic of the bulk condensate structure are obtained. Such a highly distorted transition-layer state has been well recognized during the epitaxial growth of several solids and has been analyzed theoretically by using the dislocation model.[45]

The experimental recognition that the condensed vapors have structure (lattice parameter, rotational symmetry, etc.) closely related to the substrate surface structure allows the preparation of thin films with unusual structure and electronic properties. It is hoped that this phenomenon will be explored further and utilized in studies of surface catalysis, among others.

There is a notable exception to the similar unit-cell rule that should be mentioned here. If the adsorbed layer shows partial ionization, mutual repulsion may lead to adsorption in a disordered structure. Such a system appears to be adsorbed sodium on tungsten surfaces.[46]

5.6.4. CLASSES OF CHEMISORBED SURFACE STRUCTURES

During chemisorption the gas layer adsorbed on the surface may retain its structure—similar to that it had in the gas phase. Frequently, however, it undergoes chemical changes or it induces chemical changes in the substrate on which chemisorption has taken place. Depending on the nature of the interaction of the chemisorbed layer with the substrate, it is possible to classify the different chemisorbed structures into well-defined types identifiable from the experimental data.

Chemisorption "on top." Gases may chemisorb on the surface and arrange themselves in different surface structures according to the rules of ordering given in Section 5.6.3. Adsorption "on top" implies that the chemisorbed reactants or, if they dissociate, the products of dissociation, stay on the surface and will not subsequently diffuse into the bulk to participate in bulk chemical processes. The structure that forms is a two-dimensional arrangement in which the participating atoms or molecules are those of the adsorbed gas and does not include substrate atoms to any large extent. The adsorption of olefins on platinum surfaces or the adsorption of saturated hydrocarbons on nickel surfaces provide good examples for this type of chemisorption.

Most of the adsorption studies employed ambient pressures between 10^{-9} and 10^{-4} torr because of experimental limitations. It is likely that as the pressure is increased, the larger surface coverages that can be produced may give rise to on-top surface structures that have not been able to form at these low experimental pressures. Since many of the chemisorption and surface reaction studies are carried out at high pressures as compared with those used in LEED studies, they may not be directly correlated with such studies. Efforts should be extended to establish a pressure region where low- and high-pressure studies would overlap so that one could extrapolate with confidence the results of low-pressure LEED experiments, which provide surface structural information, to high pressures. LEED studies should be extended to pressures as high as experimentally feasible. In this way the relationship between the structure of the chemisorbed layer and its reactivity can be established.

Coadsorbed surface structures. LEED studies uncovered some surface structures that would form during the simultaneous adsorption (co-adsorption) of two gases but would not form during the adsorption of only one or the other gas component. The formation of these mixed surface structures seems to be a general property of adsorbed gas layers on tungsten surfaces. It was shown that the simultaneous chemisorption of nitrogen and carbon monoxide on the (100) surface of tungsten gives a series of surface structures not all of which can be formed by the individual gases. Similar results can be obtained by the coadsorption of oxygen and carbon monoxide on tungsten (110) faces or hydrogen and carbon monoxide on the (100) surfaces of platinum (Table 5.4). The appearance of such surface structures indicates that there is a strong attractive interaction within the adsorbed layers between the unlike molecules, which arrange themselves in a mixed structure where both molecules appear to participate in the surface unit cell. These coadsorbed structures appear most frequently when both gases being adsorbed have approximately equal probability of accommodation. If one gas adsorbs much more strongly than the other (e.g., during the coadsorption of xenon and carbon monoxide on a metal surface), then one finds that the more tenacious species (carbon monoxide) will replace and displace the other species (xenon) adsorbed on the surfaces. In this case the co-adsorbed structures are unlikely to form. In fact, the surface structure of several molecules could not be studied by using LEED because of competition for the adsorption sites on the surfaces by gas atoms in the ambient, which would chemisorb preferentially. For example, carbon monoxide, which is one of the major constituents of the ambient, adsorbs preferentially on several metal surfaces and has prevented the study of the chemisorption of saturated hydrocarbons at the low pressures used in most LEED experiments.

Coadsorption and the formation of mixed surface structures indicate that in many surface chemical reaction studies the method used to introduce the different reactive gases into the chemical reaction is important. When one gas is preadsorbed on the surface and the other gas is allowed to react with the chemisorbed species, one might find different chemical reaction rates and reaction products than when the two gases are introduced simultaneously onto the surface.

Reconstructed surface structures. It has been reported from several studies that a strongly exothermic surface reaction, such as the chemi-sorption of oxygen on nickel or on other metal surfaces, can dislodge the substrate atoms from their equilibrium positions and cause rearrange-ments of the substrate structure, which is commonly called *reconstruc-tion*. The reconstructed surface structure is composed of both metal and chemisorbed atoms in periodic arrays. Although changes in the diffrac-tion pattern during chemisorption can be analyzed in several different ways, complementary experimental evidences seem to indicate that reconstruction is the most likely interpretation of the structural changes observed during the oxidation of many metal surfaces. Reconstruction can be easily rationalized by comparing the heat of adsorption of strongly chemisorbed species with the lattice energies of the different substrate metals and semiconductors. If the heats of adsorption and the lattice energies are of the same magnitude, it indicates the formation of chemical bonds between the adsorbed atom and the substrate that are similar in strength to those between the substrate atoms. Reconstruction of the surface may be looked upon as a precursor for oxidation reactions or other chemical reactions that proceed into the bulk: for example, carbide formation via carbon diffusion, or nitridation via nitrogen diffusion into the bulk by a diffusion-controlled mechanism. Since reconstruction displaces and rearranges metal atoms on the surface while forming ordered surface structures, these structures may be stable to much higher temperatures than two-dimensional surface structures which are solely due to adsorbed gases. The types of surface structures that form depend on the structure of the substrate and on the surface density of chemisorbed atoms. For example, during the initial stages of chemisorption of oxygen on the nickel (110) face, (2 × 1) and (3 × 1) surface structures are formed. The heating of these surface structures in vacuum causes their disappearance, which indicates that diffusion of oxygen from these surface structures into the bulk has occurred. Further oxygen dosing of surfaces at high temperature re-forms these surface structures, which appear to be surface intermediates during the dissolution of oxygen in the bulk nickel lattice. The dissolution of oxygen via the oxygen surface structures continues until the solubility limit of oxygen in the metal crystal is reached. At that point the metal oxide may precipitate out as a second phase at the surface. The formation of the second phase is accompanied by the appearance of streaking in the surface diffraction pattern and then the gradual appearance of new diffraction features that can be attributed to the newly formed oxide. Although reconstructed surfaces may persist to higher temperatures than those due to adsorbed gases only on top of the surface, they can often be removed by well-chosen surface chemical reactions. For example, oxide structures or structures due to chemisorbed oxygen could be removed by heating in hydrogen. Ion bombardment or high-temperature heat treatment in vacuum, which causes the vaporization of the topmost atomic layers, can also be used to restore the surface to its original unreconstructed state.

Surface reconstruction processes discovered by LEED studies give us a new view of the mechanism of chemisorption. Reconstructed surfaces may well be active surface structures in many exothermic catalytic surface reactions.

Amorphous surface structures. It has been found during studies of the adsorption of oxygen on some metal surfaces that chemisorption takes place via the formation of disordered layers. For example, the chemisorption of oxygen on aluminum surfaces or oxygen on chromium surfaces takes place in such a manner. The adsorption of carbon monoxide on the (100) faces of tantalum is another example of this type of adsorption. When the chemisorbed disordered oxygen layer is heated, oxygen from the aluminum surface diffuses into the bulk and the surface returns to its original clean, ordered, metallic state. Further dosing with oxygen at high temperatures increases the concentration of oxygen in the bulk of the metal, but the surface structure remains that of clean aluminum. This is in contrast with the behavior of oxygen on nickel or tungsten surfaces. Once the bulk of the aluminum crystal is saturated with oxygen, the surface finally loses its ordered aluminum structure and forms a disordered oxide, which now can no longer be removed by heat treatment. Under high-temperature heat treatment, in some cases there is a degree of ordering that may be taking place on the surface. However, the oxide that appears on the surface of aluminum at room temperature is characterized by the lack of ordering. Although the experimental information presently available is scanty, it appears that those oxide layers that form nonporous resistant surface films form disordered surface structures. The lack of crystallinity of the surface structures may, in future studies, be correlated with the degree of nonporosity of the deposited oxides.

In some cases heating of the adsorbed disordered layer may result in partial or complete ordering. For example, ammonia adsorbs on the (100) surface of tungsten in a disordered manner at room temperature. Upon heating to elevated temperature, a c(2 × 2) surface structure forms with the evolution of hydrogen, indicating that this structure consists of NH_2 groups adsorbed on the tungsten surface. Upon further heating the structure is rearranged into a (1 × 1)-NH_2 surface structure. Carbon monoxide seems to chemisorb at room temperature on several crystal surfaces in a disordered manner. Heating increases the surface order and aids the formation of ordered surface structures, indicating that the formation of these surface structures requires surface diffusion to occur. Therefore, in chemisorption studies, sufficient attention should be given to the thermal history and the thermal treatment being carried out after adsorption has taken place.

Three-dimensional structures. As already discussed, reconstruction of the solid surface may occur during the chemisorption of gases that induce exothermic chemical reactions at the surfaces. Reconstruction may be followed by further chemical reactions, which take place in the bulk of the solid. As the surface species diffuse into the bulk, the chemical reaction is no longer two dimensional but actually involves the species below the surface. In the final stages of oxidation when the second phase (for example, nickel oxide) is beginning to precipitate, other surface structures may appear that are characteristic of that of the bulk oxide or some mixture of the metal and the oxide structures. Three-dimensional structures also form during the carburization of tungsten. Methane decomposition yields a layer of carbon on tungsten surfaces that subsequently diffuses into the bulk. There are ordered structures at the surface during this process, in which the surface unit cells are of some integral multiple of the bulk tungsten unit cell. That is, the body-centered cubic tungsten structure appears to be maintained during the carbon diffusion process. The surface structures change from one ordered structure to another during carbon diffusion. Finally, the structure indicating the precipitation and formation of tungsten carbide, W_2C, appears at the surface.[47] Although LEED studies give information about the structure of the surface or maybe structures a few atomic layers deep at the surface, there is little doubt that these oxide or carbide structures are three dimensional. The condensation of second phases can conveniently be followed by low-energy electron diffraction because of the streaking of the diffraction pattern by the strain introduced by the phase transformation. Such studies provide us with new information about the formation of bulk phases or bulk phase transformations.

The growth of thin films of iron oxide was studied on a clean Fe(110) surface. When the iron surface was exposed to oxygen at room temperature, several surface structures were formed and then the development of a thin film of FeO(111) was observed.[48] TaO(111) has been observed to grow on the (110) face of tantalum metal after exposure to oxygen at room temperature and at high temperatures.[49] Studies of the rate of oxidation give an activation energy for the oxidation process of 0.24 eV. The growth of nickel oxide was studied when (100), (110), and (111) faces of nickel were oxidized in 10^{-6} torr of oxygen at around 500°C. On all three nickel surfaces the (100) face of the cubic oxide was formed.[50]

5.6.5. ADSORPTION ISOTHERMS OF CHEMISORBED GASES

During the chemisorption of gases, monolayer coverages are obtained at very small pressures ($\leqslant 10^{-6}$ torr) which, until recently, were outside the pressure range of convenient and reliable experimental adsorption studies. Therefore, most of the isotherms for chemisorbed gases are determined at high coverages and are often incomplete.[51] Some of the studied gas–solid systems obeyed the Langmuir isotherm (e.g., the chemisorption of H_2 on Cu), but many chemisorption processes did not. Studies of the temperature dependence of the coverage revealed that the differential heats of adsorption decrease with increasing surface coverage, nearly linearly in some cases. The differential heats of chemisorption of hydrogen on several metals, plotted in Fig. 5.19, clearly indicate the variation. Since the constancy of the heat of adsorption with coverage is one of the major assumptions in deriving the Langmuir isotherm, this simple isotherm would not be applicable under these conditions. In most chemisorption processes the heat of adsorption is thought to decrease with coverage because of the repulsive interaction between adsorbate molecules when they are in close proximity to each other. Let us assume that the heat of chemisorption ΔH_{ads}^{diff} decreases as the linear function of the coverage, which can be expressed as

$$\Delta H_{ads}^{diff} = \Delta H_{ads}(\Theta = 0)(1 - a\Theta) \qquad (5.50)$$

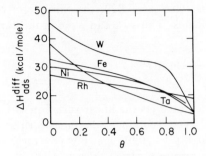

Figure 5.19. *Isosteric heats of chemisorption of hydrogen as a function of degree of covering on several metal surfaces.*

where $\Delta H_{ads}(\Theta = 0)$ is the heat of chemisorption at zero coverage and a is a proportionality constant. We will now proceed to modify the Langmuir isotherm for this condition. Equation (5.37) can be rewritten in the form

$$\frac{\Theta}{1 - \Theta} = \frac{1}{\sigma_0} F\tau \tag{5.51}$$

or, from Eq. (5.35),

$$\frac{\Theta}{1 - \Theta} = AP \exp\left(\frac{\Delta H_{ads}}{RT}\right) \tag{5.52}$$

where $A = N_A(2\pi MRT)^{-1/2}\tau_0\sigma_0^{-1}$. Substitution of Eq. (5.50) into (5.52) yields

$$\frac{\Theta}{1 - \Theta} = AP \exp\left[\frac{\Delta H_{ads}(\Theta = 0)(1 - a\Theta)}{RT}\right] \tag{5.53}$$

or, in logarithmic form after rearrangement,

$$\ln P = \ln \frac{\Theta}{1 - \Theta} - \ln A_0 + \frac{\Delta H_{ads}(\Theta = 0)a\Theta}{RT} \tag{5.54}$$

where $A_0 = A \exp\left[\Delta H_{ads}(\Theta = 0)/RT\right]$ is independent of coverage. In the range $\Theta = 0.1\text{–}0.9$, which is most accessible to chemisorption experiments, the term $\ln\left[\Theta/(1 - \Theta)\right]$ varies very slowly with coverage (it is zero at $\Theta = 0.5$). For chemisorption characterized by a strong interaction potential ($\Delta H_{ads} \gg RT$), the third term on the right side of Eq. (5.54) dominates. Thus, we can write in this range of coverages

$$\Theta \approx \frac{RT}{\Delta H_{ads}(\Theta = 0)a} \ln A_0 P \qquad (5.55)$$

This isotherm, which predicts the linear variation of Θ with $\ln P$, is the *Temkin isotherm*.[52] The Temkin isotherm as written in Eq. (5.55) is not applicable at high ($\Theta > 0.9$) or at low coverages ($\Theta < 0.1$) even if the heat of chemisorption varies linearly with the coverage, since $\ln [\Theta/(1 - \Theta)]$ is no longer negligible. However, in the middle ranges of surface coverage, this isotherm describes very well the chemisorption of many gases (H_2 on W, N_2, H_2 on Fe, etc.).

Another frequently used isotherm, the *Fraundlich isotherm*,[53] implies that the heat of chemisorption falls logarithmically with increasing surface coverage [$\Delta H_{ads}^{diff} \approx -\Delta H_{ads}(\Theta = 0) \ln \Theta$]. It is of the form

$$\sigma = cP^{1/n}$$

where c and n are constants that depend on T; n is always greater than unity and is characteristic of a given gas–surface system. Although it was first suggested as an empirical two-parameter isotherm, it can be obtained by a rigorous calculation of chemisorption on a heterogeneous surface for certain distributions of the different types of adsorption sites.[54,55] It should be noted that the predicted logarithmic $\Delta H_{ads}^{diff} - \Theta$ relationship cannot be obeyed as Θ approaches zero since ΔH_{ads} may never approach infinity at low coverages.

5.6.6. HEATS OF CHEMISORPTION

The values obtained for the heats of chemisorption are, in general, less reliable than those obtained for the heats of physical adsorption. Since large coverages can be obtained even at very low equilibrium pressures, it is often very difficult to determine the initial heat of chemisorption [$\Delta H_{ads}(\Theta \to 0)$]. In addition, because of the strong adsorbate–substrate and adsorbate–adsorbate interactions, the heats of chemisorption depend markedly on the surface structure of the substrate (distribution of different surface sites) and on the surface coverage. Heats of chemisorption vary widely between 15 and 200 kcal/mole, whereas heats of physical adsorption are all in the range 2–15 kcal/mole. Just as for physical adsorption, calorimetric and volumetric studies and other techniques may yield both the integral and the isosteric heats of chemisorption. The isosteric heat ΔH_{ads}^{diff} as a function of coverage Θ provides one with more detailed information about the nature of chemisorption. The integral heats, which are averages for the whole chemisorption process, are generally lower than the isosteric heats at $\Theta \rightarrow 0$, since the heats of chemisorption decrease with increasing coverage. In Table 5.8

Table 5.8. *Heats of Chemisorption of* O_2, H_2, N_2, *and* CO *on Several Metal Surfaces*

Gas	Material	ΔH_{ads}(kcal/mole)
O_2	W	194
	Mo	172
	Rh	118
	Pd	67
	Pt	70
H_2	Ta	45
	W	45
	Cr	45
	Mo	40
	Ni	30
	Fe	32
	Rh	28
	Pd	26
	Mn	17
N_2	W	95
	Ta	140
	Fe	70
CO	Ti	153
	W	82
	Ni	42
	Fe	46

the integral heats of adsorption are listed for O_2, H_2, CO, and N_2 on several metal surfaces.

5.6.7. FLASH DESORPTION

One of the more frequently used techniques to determine the surface concentration of adsorbed molecules and their heats of adsorption involves the monitoring of the pressure change during desorption of molecules from a heated surface. In a typical desorption experiment the sample is heated rapidly (10^2–10^3 deg/sec) and the pressure in the reaction chamber of volume V is monitored as a function of time.[153] Assuming that during the flash desorption no readsorption occurs on the heated sample, the desorption rate, $F(t)$ (molecules/cm^2 sec) of the molecules from the surface of area $\mathscr{A}$ is given by[154]

$$\frac{\mathscr{A}}{kV} F(t) = \frac{P - P_0}{\tau_c} + \frac{dP}{dt} \tag{5.56}$$

where P is the pressure rise from the steady-state pressure P_0 of the system prior to flashing, τ_c is the mean residence time of the gas atoms in the cell ($\tau_c = $ volume/pumping speed), and k is a constant. The term on the left side of Eq. (5.56) is due to the flash desorption of molecules. The first term on the right of Eq. (5.56) is the pumping rate which removes gases from the system at pressure $P > P_0$, and the second term on the right is the first derivative of the pressure with time. For high pumping speeds that remove the desorbed gas molecules rapidly from the system ($\tau_c \to 0$), Eq. (5.56) reduces to $(\mathscr{A}/kV)F(t) \approx (P - P_0)/\tau_c$; i.e., the desorption rate is proportional to the pressure change. For small pumping speeds ($\tau_c \to \infty$), a condition that can be approximated if the duration of the flash is short compared to τ_c, the desorption rate is proportional to dP/dt. The desorption rate as a function of temperature may be measured directly if one of these conditions can be approximated experimentally. From the data, the activation energy of desorption, ΔH_{des}^*, is calculated. Perhaps the most significant result of flash desorption studies was the demonstration that one type of molecule can have several discrete activation energies of desorption. This indicates the presence of discrete binding states at the surface that hold the adsorbed molecules with different binding energies.[155] Often there is evidence of interconversion between molecules in the different states.[156] For other adsorbed gases with multiple binding states there is no evidence for equilibrium among the various states. There may be multiple binding states on one crystal face while only one binding state for the same molecule on another crystal face.[99,100]

Flash desorption studies can also be used to measure surface coverages of adsorbed molecules. In addition, the sticking probabilities of gases can be accurately estimated from the desorption rates when they are monitored as a function of time.[156]

5.6.8. INTERACTION OF ELECTRONS AND ELECTROMAGNETIC RADIATION WITH ADSORBED GASES

Under bombardment by electrons of relatively low energy (<100 eV), atoms or molecules in the adsorbed gas layer can undergo various changes of state. Since the energy of the bombarding particles is much greater than the energy of the chemical bond between the adsorbate and the substrate, the bond may be broken and the molecule is desorbed. The desorption probabilities of several molecules under electron impact have been measured.[166] It was found to be high for carbon monoxide and orders of magnitude lower for oxygen from a tungsten surface. The adsorbed molecule may be ionized and subsequently desorbed.[167] The ion current obtained this way can be readily monitored. In case of multiple binding states, conversion of the chemisorbed species from one binding state can take place under electron impact.[168] The bound electrons in the adsorbed molecule can be excited to higher-energy bound electronic states that cause the absorption of the incident electron beam at energies that are characteristic of the electronic structure of the adsorbed molecule. Electron absorption studies can yield valuable information about the molecular structure of the adsorbate and promises to become an important technique of surface science.[169] Various chemical reactions can also be induced by electron impact, including dissociation[170] and polymerization[171] of the chemisorbed molecules.

Electromagnetic radiation may also be used to learn about the structure of adsorbed molecules. Infrared spectroscopy studies of the vibrational modes of adsorbed molecules are frequently carried out and often give rigorous proof that the species chemisorbed in the molecular form.[172] The major difficulty in carrying out optical studies in the adsorbed layer is to obtain sufficient intensities of the absorption peaks. For this purpose high-surface-area polycrystalline samples are often used and multiple scattering techniques have been developed.[173] Electron spin resonance can be used to detect minute concentrations of adsorbed species that have unpaired electrons.[174,175]

5.6.9. CHANGE OF WORK FUNCTION DURING ADSORPTION ON METAL SURFACES

The work function and its small variation from crystal face to crystal face due to the redistribution of electron density in the surface layer was discussed in Chapter 4. It can be readily seen that the adsorption of atoms or molecules that cause a further redistribution of the charge density at the surface gives rise to changes of the work function.[54] If the adsorbed atoms are ionized and transfer electrons into the solid surface, the work function decreases. Conversely, the formation of adsorbed negative ions increases the work function. Most frequently the adsorbed atoms or molecules are only polarized by the attractive surface forces, and then they may be viewed as dipoles aligned perpendicular to the surface.[55,56] If the adsorbed layer has its negative pole outward, the work function of the electrons in the solid (generally metal) surface increases because of the lowering of the free electron density by charge transfer and charge localization in the adsorbed layer. If the positive pole of the polarized adsorbed layer is outward, the work function decreases as a result of the excess electron density transferred to the solid.

Thus measurement of work-function changes on well-defined metal surfaces can give a great deal of information about the nature of adsorption and the type of surface bond that forms between the adsorbed molecule and the surface atoms. In some cases where the charge transfer per surface bond is additive, changes of work function during the adsorption, as a function of time, may be related to the amount of adsorbed atoms on the surface, i.e., give an estimate of relative surface coverage. In addition, competing chemisorption processes may be followed by monitoring the change in work function, especially if one of the gases acts as an electron donor and the other as an acceptor, on the solid surface.

Work-function changes upon chemisorption are measured by a variety of methods ranging from field emission[57] and retarding potential[58] to the vibrating capacitor[59] method. Since one is seeking to determine a *relative* change in the electron emission characteristics of the surface, variation of many different electrical properties of the surface upon chemisorption, such as thermionic or field emission, surface capacitance or conductance, can provide one with information about the work-function change.

Table 5.9 lists the work-function changes $\Delta\phi$ upon the adsorption of various gases on different metal surfaces; the $\Delta\phi$ values correspond to those reported at the termination of chemisorption. Nevertheless, the surface coverages, although considered complete, are not known accurately for most cases. Most of the work-function studies were carried out on polycrystalline surfaces and can be taken as averages of work-function values for the different low-index crystal faces. It can be seen from work-function changes on single-crystal surfaces (wherever reported) that variations of the ϕ values from face to face can be large because of differences in coverages, surface structure, and sometimes the type of interaction that takes place.

There are several trends in the work-function data of Table 5.9 that should be noted. Both hydrogen and oxygen adsorption on metal surfaces appears to increase the work function. The chemisorption of π-bonded organic molecules (olefins, for example) decreases the work function. Carbon monoxide, on the other hand, increases the work function of some metals (Fe, Ni, Co, W, Pt) but decreases the work function of the others (Cu, Au, Ag).

It should be mentioned that the adsorption of rare gases, such as Kr, Ar, and Xe, decreases the work function of various metals. The magnitude of this induced-dipole interaction is proportional to the polarizability of the gas atom.

Table 5.9. *Work-function Change of Metals upon Chemisorption of Several Gases*

Metal	Adsorbing gas	Work function change (eV)[58,59]
W	H_2	0.48
Fe	H_2	0.45
Ni	H_2	0.35
Cu	H_2	0.35
Ag	H_2	0.35
Au	H_2	0.18
Pt	H_2	0.14
W	O_2	1.19
Ni	O_2	1.6
Pt	O_2	1.2
Cu	O_2	0.68
Fe	CO	—1.5
Co	CO	—1.48
Ni	CO	—1.35
W	CO	—0.86
Pt	CO	0.18
Cu	CO	0.3
Ag	CO	0.31
Au	CO	0.92
Pt	C_2H_2	—1.4
Pt	C_2H_4	—1.11
Pt	C_3H_6	—1.36

5.6.10. CHARGE TRANSFER AT SEMICONDUCTOR SURFACES DURING CHEMISORPTION

Chemisorption, which always involves charge transfer between the adsorbate and the surface, can readily yield monolayer coverages on a metal surface because of the presence of mobile charge carriers in large concentrations (about one free electron per surface atom). Chemisorption on semiconductor or insulator surfaces is more difficult, however, because of the scarcity of free electrons (or holes) available for charge transfer (less than one electron per 10^6 surface atoms). As soon as all the mobile charges are transferred from the topmost atomic layer at the surface to the adsorbed gas atoms (for chemisorption involving electron donation), a space charge builds up at the surface that opposes the further transfer of charges to the adsorbed gas layer. Thus chemisorption becomes limited by the rate of charge transfer to the surface over the surface space-charge potential-energy barrier V_s. This barrier does not maintain constant height throughout the chemisorption process but increases with increased number of ionized donors (or acceptors) as $V_s \propto N_D^+$. The barrier height is also temperature dependent and varies as $\exp(-eV_s/k_BT)$.

Let us consider a typical charge-transfer-limited chemisorption of an atom A on the surface of an n-type semiconductor, which combines with an electron supplied by the bulk of the material:

$$A + e^- \rightarrow A^-$$

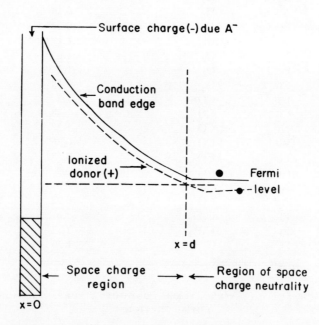

Figure 5.20. *Scheme of an electron-deficient (depletion) space-charge region at the surface of an n-type semiconductor.*

As the chemisorption proceeds, a dipole layer is formed that consists of a negatively charged chemisorbed layer (due to the presence of A^-) and a positively charged space-charge region near the surface (due to ionized donors N_D^+). This situation is depicted schematically in Fig. 5.20.

The arrival rate, dn_s/dt, of electrons at the surface, $x = 0$, by flow over the top of the potential-energy barrier V_s is given by[60]

$$\frac{dn_s}{dt} = N\left(\frac{k_BT}{2\pi m_e}\right)^{1/2} e^{-eV_s/k_BT} \tag{5.57}$$

For space-charge barriers of finite width, d, the potential, $\mathbf{V}(x)$, resulting from the dipole layer is given by Eq. (4.49) as

$$\mathbf{V}(x) = \frac{1}{2}\frac{e}{\epsilon\epsilon_0} N_D^+ (x - d)^2 \tag{5.58}$$

from which

$$\mathbf{V}_s = \frac{1}{2}\frac{e}{\epsilon\epsilon_0} N_D^+(\text{impurity})d^2 \tag{5.59}$$

since $\mathbf{V}(x = 0) = \mathbf{V}_s$. The width of the space charge can be expressed in terms of the volume density of ionized donors and the surface density of electrons, n_s, transferred to the chemisorbed layer to form A^-, as $n_s = N_D^+ d$. The height of the space-charge layer can thus be expressed as

$$\mathbf{V}_s = \frac{1}{2}\frac{e}{\epsilon\epsilon_0}\frac{n_s^2}{N_D^+} \tag{5.60}$$

The rate of charge transfer over the top of the barrier can be expressed after the substitution of Eq. (5.60) into (5.57) as

$$\frac{dn_s}{dt} = Ke^{-\alpha n_s^2} \tag{5.61}$$

where $K = N(k_B T/2\pi m_e)^{1/2}$ and $\alpha = \frac{1}{2}(e^2/\epsilon\epsilon_0)(1/N_D^+)$. Thus the more electrons are transferred to the surface into the chemisorbed layer, the lower the rate of further electron transfer from the bulk to the surface.

As soon as chemisorption commences, the surface charge builds up so rapidly that the initial rate of uptake is not likely to be measurable. Once the surface density of trapped electrons reaches a certain value, n_s^0, further electron transfer from the bulk to the surface proceeds over the space charge barrier at a measurable rate. Equation (5.61) can be integrated analytically in the region that is experimentally attainable, from $n_s = n_s^0$ at time $t = t_0$, where n_s^0 is a large surface charge already present to $n_s = n_s^0 + n$ at time t:

$$K(t - t_0) = \int_{n_s^0}^{n_s^0 + n} dn_s e^{\alpha n_s^2}$$

$$K\Delta t \approx \frac{e^{\alpha(n_s^0)^2}}{2\alpha n_s^0}(e^{2\alpha n_s^0 \cdot n} - 1)$$

where we have used the fact that $n_s^0 \gg n$ and have neglected the quadratic terms n^2 in the exponential. After rearrangement, this expression becomes

$$n = \frac{1}{2\alpha n_s^0} [\ln 2\alpha n_s^0 K e^{-\alpha(n_s^0)^2} \cdot \Delta t + 1] \tag{5.62}$$

When $2\alpha n_s^0 K e^{-\alpha(n_s^0)^2} \Delta t \gg 1$, the second term in the logarithm can be neglected and we obtain the approximate time dependence of the form

$$n \approx \frac{1}{2\alpha n_s^0} \ln \Delta t + C \tag{5.63}$$

In most experiments in which the surface conductance or the conductivity of a thin film is measured during chemisorption as a function of time, it is *not* the time dependent change of the surface density of electrons $n_s(t)$ which is transferred into the chemisorbed layer that is monitored, but the concentration of free electrons n_e which is still available for conduction. Since $n_s = N' - n_e$, where N' is the total concentration of electrons (free and trapped in chemisorbed layer), experimental plots of the current versus the logarithm of time give straight lines, indicating the relationship

$$-bn_e = \log t + C' \tag{5.64}$$

The type of rate law given by Eq. (5.63) or (5.64) is frequently called the *Elovich rate equation*.[61] The chemisorption of oxygen on several semiconductor surfaces (Ge, ZnO, CdSe) obeys this rate law.[62,63] The oxygen uptake rate is exponentially dependent upon the charge concentration at the surface. Frequently a fast pressure-independent electron-transfer process takes place first, followed by a slower pressure-dependent chemisorption. Both processes appear to be largely irreversible and follow the rate equation of the form (5.64). The slower pressure-dependent chemisorption rate may be expressed as

$$-\frac{dn_e}{dt} = P_{O_2}^{1/n} K e^{\alpha' n_e^2} \tag{5.65}$$

where n was found to be 2 or 4. The pressure dependence, which appears at the latter stages of chemisorption, may be due to the poor accommodation of oxygen on the partially oxygen-covered semiconductor surface, which lowers the surface concentration of adsorbed species available for charge transfer. Typical conductance curves as a function of the logarithm of time for oxygen chemisorption on CdSe surfaces are shown in Fig. 5.21 for several oxygen pressures at 360°C.

According to the "boundary-layer" theory[64,65] in any surface reaction where electron transfer through a potential barrier is the rate-determining step, one expects the rate to be exponentially dependent on the transferred

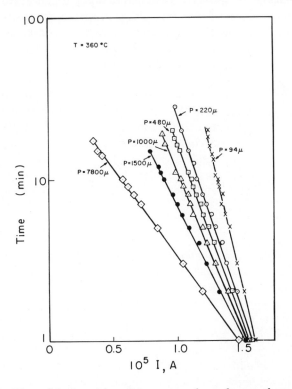

Figure 5.21. *Plots of the logarithm of time versus the surface conductance change during the chemisorption of oxygen on the surface of cadmium selenide at various oxygen pressure.*

electron concentration to give the type of rate law which was obtained in Eq. (5.57) or (5.65). This treatment has been expanded for space-charge layers and chemisorption processes of many types. It should hold even for electron tunneling through a thin potential barrier.

Any adsorption mechanism that requires an activation energy ΔE, and for which the activation energy increases with increasing coverage Θ ($\Delta E = b\Theta$), may yield a rate equation of the form of Eq. (5.61): $d\Theta/dt \propto Ae^{-b\Theta/RT}$. Thus it is not surprising that the Elovich equation describes the adsorption rate of different gases on a wide variety of solid surfaces.

The importance of the carrier concentration and the type of carriers available at the surface for chemical processes on semiconductor or insulator surfaces is indicated by several studies. The activation energy of chemisorption can be changed markedly by using *n*-type, instead of *p*-type, semiconductors.[66] Even the nature of the chemical reaction may change upon changes of the space-charge properties and carrier concentrations. For example, the oxidation of propylene to acrolein on cuprous oxide crystal surfaces was found to depend on the mobile-hole concentration at the surface, which could be changed by variation of the partial pressure of oxygen over the solid.[67] Low oxygen partial pressures resulted in copper-rich cuprous oxide surfaces that exhibited low hole concentrations. Increase of the oxygen partial pressure gave an oxygen-rich—Cu_2O surface with increased hole conductivity Thus the height of the space-charge barrier could be altered markedly by variation of the oxygen pressure. The acrolein yield was the highest for nearly stoichiometric Cu_2O, while the oxygen-rich (strongly *p*-type) cuprous oxide surfaces favored complete oxidation of propylene to carbon dioxide and water.

5.6.11. THE SURFACE CHEMICAL BOND

There have been several attempts to correlate the experimental data on heats of chemisorption, changes of work function upon adsorption on metal surfaces, and charge transfer at semiconductor surfaces, in order to develop a theory of the surface chemical bond. In spite of the experimental uncertainties of the values obtained in many of these studies, some of the models proposed could explain qualitatively many of the experimental trends. Qualitatively three types of adsorption are distinguished, depending on the magnitude of the effective electronic charge that is localized in the neighborhood of the adsorbate.[157] *Ionic adsorption* occurs upon the transfer of an electron from the adsorbed atom to the substrate (or vice versa). The heats of chemisorption of alkali metals could be well approximated by calculations assuming ionic bond between the substrate metal and the adsorbed alkali metal atoms.[158] The agreement between measured and calculated values appears to be satisfactory (see Table 5.10).

During *neutral adsorption* there is virtually no charge transfer between the adsorbate and substrate atoms. The interaction of rare gas atoms with metal or insulator surfaces can be explained adequately by considering the effect of attractive surface forces of different types ($V \propto -r^{-3}$, $V \propto -r^{-6}$) on the approaching atom. Calculations using the interaction potentials which were given in Eqs. (5.3), (5.4), and (5.6)

Table 5.10. *Experimental and Calculated Initial Heats of Chemisorption of* Na, K, and Cs *on Tungsten*

	$\Delta H_{ads}(\Theta = 0)_{exptl}$ (kcal/mole)	$\Delta H_{ads}(\Theta = 0)_{calc}$ (kcal/mole)
Na on W	32.0	30.5
K on W	—	40.3
Cs on W	64.0	45.7

give heats of adsorption which are of the same magnitude as those measured in adsorption studies for rare gases (with the exception of xenon).

The chemisorption of most atoms and molecules represents the intermediate case where only part of an electronic charge has been transferred between the adsorbate and the substrate. If chemisorption takes place on a metal surface this case is called *metallic adsorption*. Theoretical calculations[158] using simple models have been able to estimate the effective charge on the adsorbate atoms for selected adsorbate–metal systems.[159] Partial charge transfer takes place in *chemisorption on semiconductor and insulator surfaces* as well. The nature of this type of interaction has also been the subject of several theoretical investigations.[160,161]

Several trends have been noted when correlating the adsorption data for the same gas molecule on different solids. For example, the heats of chemisorption of C_2H_4, CO, and H_2 on different transition metals decrease with increasing concentration of d electrons, as shown in Fig. 5.22. Flash desorption studies also indicated significant differences in the binding energies of hydrogen on molybdenum and tungsten.[156a] The results could be interpreted in terms of the greater bonding capabilities of the $5d$ versus the $4d$ electrons of the substrates. Field-ion-microscopy studies of the binding of ad-atoms of many metals on tungsten surfaces revealed variation of the binding energy with electronic configuration of the ad-atom.[162a] Thus the d orbitals of metal atoms and, in general, the electron configuration of single atoms appear to play an important role in forming the surface bond.[162b]

There is a great deal of experimental evidence indicating that the strength of the surface chemical bond depends on the *structure* of the surface. The predominance of ordering during chemisorption as revealed by low-energy electron diffraction and the different rules of ordering[163] already indicate that chemisorption is sensitive to the substrate structure (rotational multiplicity, for example).

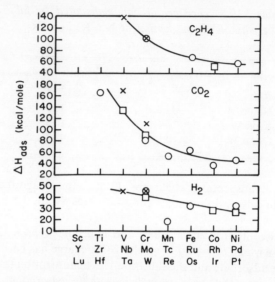

Figure 5.22. *Heats of chemisorption of ethylene, carbon dioxide, and hydrogen on various transition metal surfaces.*

Flash desorption[156] and field-ion-microscopy studies[162a] have shown that the binding energies of adsorbed atoms depend on the atomic structure of the surface and are different in most cases from crystal face to crystal face. The presence of multiple binding states could be correlated with ordered chemisorption at different crystallographically nonequivalent surface sites. Ion neutralization spectroscopy (INS) studies,[164] which detect the energy distribution of bound states of the most loosely held electrons on the adsorbed atom, showed that the bound-state energies change with changing surface structure of the chemisorbed atoms [O, S, and Se on Ni(100)]. Thus the structures of the substrate surface and of the adsorbate layer should be taken into account in developing a workable model for the surface chemical bond.

From past and recent results of chemisorption studies a consistent physical picture of chemisorption seems to emerge. Chemisorption involves charge transfer between the adsorbate and the substrate. The nature of the chemical bond of the chemisorbed gas depends partly on the electronic configuration of the *single substrate atom* and partly on the *structure of the substrate surface*. These two parameters, the electronic configuration of the single atom and the structure of the condensed phase (coordination number), have been successfully used to develop theories of chemical bonding for compounds of transition metals.[165]

A comparison of the chemistry of chemisorbed gases on metal surfaces with the chemistry of metalloorganic compounds for which the single-atom chemical properties are dominant should provide a great deal of information about that component of the surface chemical bond that is determined by interaction with a single metal atom. The success of crystal field theory in explaining certain catalytic properties of transition metals is an indication of the validity of this approach.[68] Extension of studies of the surface structures of chemisorbed gases and their variation with surface coverage and temperature should aid in exploring the structure dependence of the surface bond. Separation and identification of these two components of surface chemical interaction will lead to the development of a more complete description of the surface chemical bond.

Summary

A gas atom feels an attractive potential upon approaching the surface. The magnitude of the heat of adsorption ΔH_{ads} is indicative of the strength of the gas–surface interaction. Atomic and molecular beam scattering studies from well-defined surfaces reveal the nature of the energy transfer between the incident gas and the surface atoms. Adsorbed atoms or molecules predominantly form ordered structures on crystal surfaces. The surface structures that form depend on the symmetry of the substrate and the chemistry, size, and concentration of the adsorbed gas molecules. Measurement of the surface coverage as a function of gas pressure (adsorption isotherm) permits the determination of the thermodynamics parameters that characterize the adsorbed layer and the surface area. Changes of work function and surface conductance during adsorption provide information about charge transfer between the gas and surface atoms. The structure, the thermodynamics, and the electrical properties of the adsorbed layer reveal the nature of the surface chemical bond.

References

1. M. D. Scheer, R. Klein, and J. D. McKinley: *J. Chem. Phys.* **55**, 3577 (1971).
2a. J. H. Singleton and G. D. Halsey, Jr.: *Can. J. Chem.*, **33**, 184 (1955).
2b. W. H. Wade and L. J. Slutsky: *J. Chem. Phys.*, **40**, 3994 (1964).
3. J. E. Lennard-Jones: *Trans. Faraday Soc.*, **28**, 333 (1932).
4. J. Bardeen: *Phys. Rev.*, **58**, 727 (1940).
5. H. Morgenau and W. G. Pollard: *Phys. Rev.*, **60**, 128 (1941).

6. D. Lands and L. J. Slutsky: *J. Chem. Phys.*, **52**, 1510 (1970).
7. F. London: *Z. Physik*, **63**, 245 (1930); *Trans. Faraday Soc.*, **33**, 8 (1937).
8. P. Debye: *Phys. Z.*, **21**, 178 (1920); **22**, 302 (1921).
9. W. H. Keesom: *Phys. Z.*, **22**, 129 (1921); **22**, 643 (1921).
10. J. C. Slater and J. G. Kirkwood: *Phys. Rev.*, **37**, 682 (1931).
11. F. London: *Z. Physik. Chem.*, **11**, 222 (1930).
12. A. D. Crowell: in *Surface Forces and the Solid–Gas Interface*, E. A. Flood, ed. Marcel Dekker, Inc., New York, 1967.
13. L. T. Cowley, M. A. D. Fluendy, and K. P. Lowley: *Rev. Sci. Instr.*, **41**, 666 (1970).
14a. I. Estermann and O. Stern: *Z. Physik*, **61**, 95 (1930).
14b. J. C. Crews: *J. Chem. Phys.*, **37**, 2004 (1962).
15. N. Abuaf, J. B. Anderson, R. P. Andres, J. B. Fenn, and D. G. Marsden: *Science*, **155**, 997 (1967).
16a. R. M. Logan and R. E. Stickney: *J. Chem. Phys.*, **44**, 196 (1966).
16b. R. M. Logan, J. C. Keck, and R. E. Stickney: *Rarefied Gas Dynamics*, **1**, 49 (1967).
17. R. M. Logan and J. C. Keck: *J. Chem. Phys.*, **49**, 860 (1968).
18. R. E. Stickney: in *Structure and Chemistry of Solid Surfaces*, G. A. Somorjai, ed. John Wiley & Sons, Inc., New York, 1969.
19. F. Reif: *Statistical and Thermal Physics*. McGraw-Hill Book Company, New York, 1965.
20. F. C. Hurlbut and K. Jakus: in *Structure and Chemistry of Solid Surfaces*, G. A. Somorjai, ed. John Wiley & Sons, Inc., New York, 1969.
21. L. M. Raff, J. Lorenzen, and B. C. McCoy: *J. Chem. Phys.*, **46**, 4265 (1967).
22. B. McCarroll: *J. Chem. Phys.*, **39**, 1317 (1963).
23. J. H. deBoer: *The Dynamical Character of Adsorption*. Oxford University Press, New York, 1968.
24. J. J. Lander and J. Morrison: *Surface Sci.*, **6**, 1 (1967).
25. J. M. Dickey, H. H. Farrell, and M. Strongin: *Surface Sci.*, **23**, 448 (1970).
26. R. F. Steiger, J. M. Morabito, G. A. Somorjai, and R. H. Muller: *Surface Sci.*, **14**, 273 (1969).
27. I. Langmuir: *J. Am. Chem. Soc.*, **40**, 1361 (1918).
28. S. Brunauer, P. H. Emmett, and E. Teller: *J. Am. Chem. Soc.*, **60**, 309 (1938).
29. W. D. Harkins and G. Jura: *J. Chem. Phys.*, **12**, 113 (1944); *J. Am. Chem. Soc.*, **66**, 1366 (1944).
30. W. D. Harkins and G. Jura: *J. Am. Chem. Soc.*, **66**, 1362 (1944).
31. M. L. McClellan and H. F. Harnsberger: *J. Colloid Interface Sci.*, **23**, 577 (1967).
32a. W. M. Champion and G. D. Halsey, Jr.: *J. Am. Chem. Soc.*, **76**, 974 (1954).
32b. W. A. Steele: *Advan. Colloid Interface Sci.*, **1**, 3 (1967).
33a. D. M. Young and A. D. Crowell: *Physical Adsorption of Gases*. Butterworth & Co. (Publishers) Ltd., London, 1962.

33b. S. Ross and J. P. Oliver: *On Physical Adsorption.* John Wiley & Sons, Inc. (Interscience Division), New York, 1964.

34. S. P. Wolsky and A. B. Fowler: in *Semiconductor Surface Physics.* University of Pennsylvania Press, Philadelphia, 1957.

35. J. C. Tracy and P. W. Palmberg: *Surface Sci.*, **14**, 274 (1969).

36a. R. Defay and I. Prigogine: *Surface Tension and Adsorption.* John Wiley & Sons, Inc., New York, 1966.

36b. V. K. LaMer: *Retardation of Evaporation by Monolayers.* Academic Press, Inc., New York, 1962.

37. W. D. Harkins: *The Physical Chemistry of Surface Films.* Van Nostrand Reinhold Company, New York, 1952.

38. A. W. Adamson: *Physical Chemistry of Surfaces.* John Wiley & Sons, Inc. (Interscience Division), New York, 1967.

39. N. F. Ramsey: *Molecular Beams.* Oxford University Press, New York, 1969.

40. R. L. Palmer, H. Saltsburg, and J. N. Smith, Jr.: *J. Chem. Phys.*, **50**, 4661 (1969).

41. R. J. Maddix and A. A. Susu: *Surface Sci.*, **20**, 377 (1970).

42. R. A. Krakowski and D. R. Olander: *J. Chem. Phys.*, **49**, 5027 (1968).

43. L. A. West and G. A. Somorjai: *Proc. Intern. Conf. Solid Surf.*, *J. Vac. Sci. and Technol.* (to be published, 1972).

44. C. W. Tucker, Jr.: *J. Appl. Phys.*, **37**, 3013 (1966).

45. W. A. Jesser: *Mat. Sci. Eng.*, **4**, 279 (1969).

46. R. L. Gerlach and T. N. Rhodin: *Surface Sci.*, **17**, 32 (1969).

47. M. Boudart and D. F. Ollis: in *The Structure and Chemistry of Solid Surfaces*, G. A. Somorjai, ed. John Wiley & Sons, Inc., New York, 1969.

48. J. E. Boggio and H. E. Farnsworth: *Surface Sci.*, **3**, 62 (1964).

49. A. J. Pignocco and G. E. Pellisier: *Surface Sci.*, **7**, 261 (1967).

50. A. U. MacRae: *Science*, **139**, 379 (1963).

51. D. O. Hayward and B. M. W. Trapnell: *Chemisorption.* Butterworth & Co. (Publishers) Ltd., London, 1964.

52. S. Brunauer, K. S. Love, and R. G. Keenan: *J. Am. Chem. Soc.*, **64**, 751 (1942).

53. H. Freundlich: *Colloid and Capillary Chemistry.* Methuen & Co. Ltd., London, 1926.

54. C. Riviere: in *Solid State Surface Science.* M. Greene, ed. Marcel Dekker, Inc., New York, 1970.

55. F. C. Tompkins: in *Gas-Solid Interface*, E. A. Flood, ed. Marcel Dekker, Inc., New York, 1967.

56. R. V. Culver, J. Pritchard, and F. C. Tompkins: in *Surface Activity*, J. H. Shulman, ed. Academic Press, Inc., New York, 1958.

57. R. Gomer: *J. Chem. Phys.*, **21**, 1869 (1953).

58. B. Gysae and S. Wagener: *Z. Physik*, **115**, 296 (1940).

59. J. C. P. Mignolet: *Discussions Faraday Soc.*, **8**, 326 (1950).

60. G. A. Somorjai and R. R. Haering: *J. Phys. Chem.*, **67**, 1150 (1963).

61. S. Y. Elovich and J. Zhabrova: *J. Phys. Chem. U.S.S.R.*, **13**, 1761 (1939).

62a. S. R. Morrison: *J. Phys. Chem.*, **57**, 860 (1953).

62b. G. A. Somorjai: *J. Phys. Chem. Solids*, **24**, 175 (1963).

63. M. Green, J. A. Kafalas, and P. H. Robinson: in *Semiconductor Surface Physics.* University of Pennsylvania Press, Philadelphia, 1957.

64. K. Hauffe and H. J. Engell: *Z. Elektrochem.*, **56**, 366 (1952).

65. P. B. Weisz: *J. Chem. Phys.*, **21**, 1531 (1953).

66. A. C. Zettlemoyer and R. D. Tyengar: in *The Gas-Solid Interface*, E. A. Flood, ed. Marcel Dekker, Inc., New York, 1967.

67. B. J. Wood, H. Wise, and R. S. Yolles: *J. Catalysis*, **15**, 355 (1969).

68. G. C. Bond: *Catalysis by Metals.* Academic Press, Inc., New York, 1962.

69. A. U. MacRae: *Surface Sci.*, **1**, 319 (1964).

70. L. H. Germer, E. J. Scheibner, and C. D. Hartman: *Phil. Mag.*, **5**, 222 (1960).

71. R. L. Park and H. E. Farnsworth: *Appl. Phys. Letters*, **3**, 167 (1963).

72. T. Edmonds and R. C. Pitkethly: *Surface Sci.*, **15**, 137 (1969).

73. J. W. May and L. H. Germer: in *The Structure and Chemistry of Surfaces*, G. A. Somorjai, ed. John Wiley & Sons, Inc., New York, 1969.

74. L. H. Germer and A. U. MacRae: *J. Chem. Phys.*, **36**, 1555 (1962).

75. J. C. Bertolini and G. Dalmai-Imelik: *Rept. Inst. de Rech. sur la Catalyse —Villeurbonne*, 1969.

76. B. D. Campbell, C. A. Haque, and H. E. Farnsworth: in *The Structure and Chemistry of Solid Surfaces*, G. A. Somorjai, ed. John Wiley & Sons, Inc., New York, 1969.

77. H. E. Farnsworth and J. Tuul: *J. Phys. Chem. Solids*, **9**, 48 (1958).

78. J. W. May and L. H. Germer: *Surface Sci.*, **11**, 443 (1968).

79. H. E. Farnsworth: *Appl. Phys. Letters*, **2**, 199 (1963).

80. R. E. Schlier and H. E. Farnsworth: *Advan. Catalysis*, **9**, 434 (1957).

81. L. H. Germer and C. D. Hartman: *J. Appl. Phys.*, **31**, 2085 (1960).

82. H. E. Farnsworth and H. H. Madden, Jr.: *J. Appl. Phys.*, **32**, 1933 (1961).

83. R. L. Park and H. E. Farnsworth: *J. Chem. Phys.*, **43**, 2351 (1965).

84. L. H. Germer: *Advan. Catalysis*, **13**, 191 (1962).

85. L. H. Germer, R. Stern, and A. A. MacRae: in *Metal Surfaces.* ASM, Metals Park, Ohio, 1963, p. 287.

86. M. Orchis and H. E. Farnsworth: *Surface Sci.*, **11**, 203 (1968).

87. H. E. Farnsworth, R. E. Schlier, T. H. George, and R. M. Buerger: *J. Appl. Phys.*, **29**, 1150 (1958).

88. L. H. Germer and A. U. MacRae: *Robert A. Welch Found. Res. Bull. No. 11*, 5 (1961).

89. R. L. Park and H. E. Farnsworth: *J. Chem. Phys.*, **40**, 2354 (1964).

90. L. H. Germer, J. W. May, and R. J. Szostak: *Surface Sci.*, **7**, 430 (1967).

91. A. G. Jackson and M. P. Hooker: *Surface Sci.*, **6**, 279 (1967).

92. L. H. Germer and A. U. MacRae: *Proc. Natl. Acad. Sci. U.S.*, **48**, 997 (1962).

93. C. A. Haque and H. E. Farnsworth: *Surface Sci.*, **1**, 378 (1964).

94. C. W. Tucker, Jr.,: in *The Structure and Chemistry of Solid Surfaces*, G. A. Somorjai, ed. John Wiley & Sons, Inc., New York, 1969.

95. C. W. Tucker, Jr.: *Surface Sci.*, **2**, 516 (1964).
96. C. W. Tucker, Jr.: *Appl. Phys. Letters*, **3**, 98 (1963).
97. C. W. Tucker, Jr.: *J. Appl. Phys.*, **35**, 1897 (1964).
98. J. M. Charlot and R. Deleight: *Compt. Rend.* **259**, 2977 (1964).
99. A. E. Morgan and G. A. Somorjai: *J. Chem. Phys.*, **51**, 3309 (1969).
100. A. E. Morgan and G. A. Somorjai: *Surface Sci.*, **12**, 405 (1968).
101. A. E. Morgan and G. A. Somorjai: *Trans. Am. Cryst. Assoc.*, **4**, 59 (1968).
102. C. Burggraf and Sime Mosser: *Compt. Rend.*, **268**, 1167 (1969).
103. J. C. Tracy and P. W. Palmberg: *J. Chem. Phys.*, **51**, 4852 (1969).
104. R. O. Adams: in *The Structure and Chemistry of Solid Surfaces*, G. A. Somorjai, ed. John Wiley & Sons, Inc., New York, 1969.
105. L. Trepte, C. Menzel-kopp, and E. Menzel: *Surface Sci.*, **8**, 223 (1967).
106. R. E. Schlier and H. E. Farnsworth: *J. Appl. Phys.*, **25**, 1333 (1954).
107. G. W. Simmons, D. F. Mitchell, and K. R. Lawless: *Surface Sci.*, **8**, 130 (1967).
108. H. E. Farnsworth and D. M. Zehner: *Surface Sci.*, **17**, 7 (1969).
109. F. Jona: *J. Phys. Chem. Solids*, **28**, 2155 (1967).
110. S. M. Bedair, F. Hoffman, and H. P. Smith, Jr.: *J. Appl. Phys.*, **39**, 4026 (1968).
111. H. H. Farrell: Ph.D. Dissertation, University of California, Berkeley, Calif., 1969.
112. C. W. Tucker, Jr.: *J. Appl. Phys.*, **37**, 3013 (1966).
113. C. W. Tucker, Jr.: *J. Appl. Phys.*, **38**, 2696 (1967).
114. C. W. Tucker, Jr.: *J. Appl. Phys.*, **37**, 4147 (1966).
115. W. P. Ellis: *J. Chem. Phys.*, **48**, 5696 (1968).
116. L. H. Germer: *Phys. Today*, 19 (July 1964).
117. L. H. Germer and J. W. May: *Surface Sci.*, **4**, 452 (1966).
118. J. W. May, L. H. Germer, and C. C. Chang: *J. Chem. Phys.*, **45**, 2383 (1966).
119. J. W. May and L. H. Germer: *J. Chem. Phys.*, **44**, 2895 (1966).
120. M. Boudart and D. F. Ollis: in *The Structure and Chemistry of Solid Surfaces*, G. A. Somorjai, ed. John Wiley & Sons, Inc., New York, 1969.
121. P. J. Estrup and J. Anderson: *J. Chem. Phys.*, **49**, 523 (1968).
122. C. C. Chang and L. H. Germer: *Surface Sci.*, **8**, 115 (1967).
123. T. C. Tracy and J. M. Blakely: in *The Structure and Chemistry of Solid Surfaces*, G. A. Somorjai. ed. John Wiley & Sons, Inc., New York, 1969.
124. J. W. May, R. J. Szostak, and L. H. Germer: *Surface Sci.*, **15**, 37 (1969).
125. C. C. Chang: *J. Electrochem. Soc.*, **115**, 354 (1968).
126. P. J. Estrup: in *The Structure and Chemistry of Solid Surfaces*, G. A. Somorjai, ed John Wiley & Sons, Inc., New York, 1969.
127. P. J. Estrup and J. Anderson: *J. Chem. Phys.*, **46**, 567 (1967).
128. P. W. Tamm and L. D. Schmidt: *J. Chem. Phys.*, **51**, 5352 (1969).
129. P. J. Estrup and J. Anderson: *J. Chem. Phys.*, **45**, 2254 (1966).
130. J. E. Boggio and H. E. Farnsworth: *Surface Sci.*, **1**, 399 (1964).

131. T. W. Haas: in *The Structure and Chemistry of Solid Surfaces*, G. A. Somorjai, ed. John Wiley & Sons, Inc., New York, 1969.

132. T. W. Haas, A. G. Jackson, and M. P. Hooker: *J. Chem. Phys.*, **46**, 3025 (1967).

133. H. H. Madden and H. E. Farnsworth: *J. Chem. Phys.*, **34**, 1186 (1961).

134. K. K. Vijai and P. F. Packman: *J. Chem. Phys.*, **50**, [3] 1343 (1969).

135. A. J. Pignocco and G. E. Pellisier: *Surface Sci.*, **7**, 261 (1967).

136. K. Moliere and F. Portele: in *The Structure and Chemistry of Solid Surfaces*, G. A. Somorjai, ed. John Wiley & Sons, Inc., New York. 1969.

137. T. W. Haas and A. G. Jackson: *J. Chem. Phys.*, **44**, 2921 (1966).

138. H. E. Farnsworth and K. Hayek: *Nuovo Cimento, Suppl.*, **5**, 2 (1967).

139. K. Hayek and H. E. Farnsworth: *Surface Sci.*, **10**, 429 (1968).

140. G. J. Dooley and T. W. Haas: *J. Chem. Phys.*, **52**, 993 (1970).

141. H. K. A. Kann and S. Feuerstein: *J. Chem. Phys.*, **50**, 3618 (1969).

142. R. E. Schlier and H. E. Farnsworth: *J. Chem. Phys.*, **30**, 917 (1959).

143. J. J. Lander and J. Morrison: *J. Appl. Phys.*, **33**, 2089 (1962).

144. A. J. Van Bommel and F. Meyer: *Surface Sci.*, **6**, 39 (1967).

145. R. Heckingbottom: in *The Structure and Chemistry of Solid Surfaces*, G. A. Somorjai, ed. John Wiley & Sons, Inc., New York, 1969.

146. A. J. Van Bommel and F. Meyer: *Surface Sci.*, **8**, 381 (1967).

147. L. H. Germer and A. U. MacRae: *J. Appl. Phys.*, **33**, 2923 (1962).

148. H. E. Farnsworth, R. E. Schlier, T. H. George, and R. M. Buerger: *J. Appl. Phys.*, **29**, 1150 (1958).

149. J. J. Lander and J. Morrison: *J. Chem. Phys.*, **37**, 729 (1962).

150. J. B. Marsh and H. E. Farnsworth: *Surface Sci.*, **1**, 3 (1964).

151. D. Haneman: *Phys. Rev.*, **119**, 567 (1960).

152. J. J. Lander and J. Morrison: *J. Appl. Phys.*, **34**, 1411 (1963).

153. P. A. Redhead: *Vacuum*, **12**, 203 (1962).

154. R. Gomer: in *Gas-Surface Interactions*, H. Saltsburg et al., eds. Academic Press, Inc., New York, 1967.

155. G. Erlich: *J. Chem. Phys.*, **36**, 1171 (1962).

156a. H. R. Han and L. D. Schmidt: *J. Phys. Chem.*, **75**, 227 (1971).

156b. P. W. Tamm and L. D. Schmidt: *J. Chem. Phys.*, **54**, 4775 (1971).

156c. P. W. Tamm and L. D. Schmidt: *J. Chem. Phys.*, **52**, 1150 (1970).

157. R. Gomer and L. W. Swanson: *J. Chem. Phys.*, **38**, 1613 (1963).

158a. A. J. Bennett and L. M. Falicov: *Phys. Rev.*, **151**, 512 (1966).

158b. T. B. Grimley: *Proc. Phys. Soc.*, **92**, 776 (1967).

158c. T. B. Grimley: *J. Am. Chem. Soc.*, **90**, 3016 (1968).

158d. T. B. Grimley: *Surface Sci.*, **14**, 395 (1969).

159. L. D. Schmidt and R. Gomer: *J. Chem. Phys.*, **45**, 1605 (1966).

160. M. Green: in *Semiconductor Surface Physics*. University of Pennsylvania Press, Philadelphia, 1956.

161. M. Tomosek: *Surface Sci.*, **2**, 8 (1964).

162a. E. W. Plummer and T. N. Rhodin: *J. Chem. Phys.*, **49**, 3479 (1968).

162b. T. B. Grimley: in *Molecular Processes on Solid Surfaces*, McGraw-Hill Book Company, New York, 1968.

163. G. A. Somorjai and F. J. Szalkowski: *J. Chem. Phys.*, **54**, 389 (1971).

164. H. D. Hagstrum and G. E. Becker: in *The Structure and Chemistry of Solid Surfaces*, G. A. Somorjai, ed. John Wiley & Sons, Inc., New York, 1969.

165. L. Brewer: in *Electronic Structure and Alloy Chemistry*, P. A. Beck, ed. John Wiley & Sons, Inc. (Interscience Division), New York, 1963.

166. D. Menzel and R. Gower: *J. Chem. Phys.*, **40**, 1164 (1964).

167. D. Lichtman and R. B. McQuistan: *Surface Sci.*, **5**, 120 (1966).

168a. J. T. Yates and T. E. Madey: in *The Structure and Chemistry of Solid Surfaces*, G. A. Somorjai, ed. John Wiley & Sons, Inc., New York, 1969.

168b. R. A. Armstrong: in *The Structure and Chemistry of Solid Surfaces*, G. A. Somorjai, ed. John Wiley & Sons, Inc., New York, 1969.

169. L. M. Hunter, D. Lewis, and W. H. Hamill: *J. Chem. Phys.*, **52**, 1733 (1970).

170. J. Anderson and P. J. Estrup: *Surface Sci.*, **9**, 463 (1968).

171. T. Hayashida: *Shinku*, **10**, 319 (1967).

172. R. P. Eischens and J. Jacknow: in *IIIrd Intern. Congr. Catalysis*. North-Holland Publishing Company, Amsterdam, 1964.

173. C. H. Amberg: in *The Solid-Gas Interface*. Marcel Dekker, Inc., New York, 1966.

174. K. M. Soucier: *J. Catalysis*, **5**, 314 (1966).

175. D. J. Miller and D. Haneman: *Surface Sci.*, **19**, 45 (1969).

Problems

5.1. The average heat of adsorption of hydrogen and krypton on tungsten can be estimated to be 45 and 2 kcal/mole, respectively. Estimate the residence times of these two species at 3, 300, and 1500°K, assuming that $T_s = T_g$.

5.2. Estimate the distance of closest approach to the surface at which the $\sim r^{-3}$ and $\sim r^6$ interaction energies are no longer "distinguishable" by the incident atom or molecule.

5.3. It has been proposed that surface roughness on an atomic scale should modify the scattering distributions of atomic beams as derived by the hard-cube model [T. J. Healy in *Gas-Surface Interactions*, H. Salsburg, ed., Academic Press, Inc., New York, 1967, and L. A. West and G. A. Somorjai, *J. Chem. Phys.* **54**, 2864 (1971)]. Review these papers and discuss the possible effects of surface roughness and the magnitude of its effect.

5.4. Calculate the most probable velocity of scattered Xe, Kr, Ar, and He beams for $\Theta = 45°$ and $\phi = 45°$ and $\phi = 55°$ for $T_g = T_s = 300°K$.

5.5. Using silica gel and nitrogen as absorbent, the following adsorption isotherm data were obtained (G. Constabaris et al., Chevron Research Co.):

P/P_0	$\sigma(cc/g)$
0.055	131.3
0.061	134.3
0.077	139.9
0.094	148.9
0.120	153.5
0.158	164.0
0.177	169.3
0.209	176.9
0.240	184.5
0.270	192.3
0.300	200.0
0.330	207.7
0.352	212.7

Plot the BET function, $P/\sigma(P_0 - P)$ versus P/P_0, compute the slope and the intercept, and obtain the surface area and the parameter c.

5.6. Nitrobenzene appears to be chemisorbed in a disordered layer at 300°K on the (100) and (111) platinum crystal surfaces. Ennumerate those parameters that are characteristic of the surface which control ordering. What kind of experiments would you perform to aid ordering of the chemisorbed monolayer? [See, for example, *J. Chem. Phys.*, **54**, 389 (1971).]

5.7. Oxygen diffuses into aluminum single crystals while the surface structure remains characteristic of that of pure aluminum metal. Only after virtual saturation of oxygen in the bulk will the amorphous surface form aluminum oxide [*J. Phys. Chem. Solids*, **28**, 2155 (1967)]. Nickel forms new ordered surface structures in the presence of oxygen which are characterized by unit cells that are integral multiples of the clean nickel surface unit mesh. Again, nickel oxide precipitation occurs only after saturation of the bulk with oxygen [*Surface Sci.*, **1**, 319 (1964)]. Copper, on the other hand, shows the formation of copper oxide islands at the surface even at the initial stages of oxidation [*Surface Sci.*, **8**, 130 (1967)]. Discuss the possible reasons for the differences in the surface structural changes that occur during the oxidation of these metals and attempt to predict the surface structural changes that you would expect during the oxidation of silver, tungsten, and chromium.

Appendix

Physical Constants

1 angstrom (Å) = 10^{-8} cm

Planck's constant (h) = 6.626 × 10^{-27} erg sec

Planck's constant/2π $(\hbar)$ = 1.054 × 10^{-27} erg sec

Bohr radius of ground state of hydrogen $\dfrac{\hbar^2}{me^2}$ (a_0) = 0.529 Å

Avogadro's number (N_A) = 6.0225 × 10^{23} mole^{-1}

Gas constant (R) = 8.31 × 10^7 erg mole^{-1} deg^{-1}

$\qquad\qquad\qquad$ = 1.9872 cal deg^{-1} mole^{-1}

$\qquad\qquad\qquad$ = 8.3143 J deg^{-1} mole

$\qquad\qquad\qquad$ = 82.056 cm^3 atm deg^{-1} mole^{-1}

Boltzmann's constant (k_B) = 1.3805 × 10^{-16} erg deg^{-1}

Electronic charge (e) = 1.6021 × 10^{-19} C

$\qquad\qquad\qquad$ = 4.8030 × 10^{-10} esu

Electron rest mass (m_e) = 9.1083 × 10^{-28} g

Proton rest mass (m) = 1.67237 × 10^{-24} g

Velocity of light (c) = 2.99793 × 10^{10} cm sec^{-1}

Energy-conversion Factors

	cal/mole	J/mole	cc atm/mole	cm^{-1}	eV	erg/molecule
1 cal/mole	1	4.1840	41.292	0.34974	4.3351×10^{-5}	6.9465×10^{-17}
1 J/mole	0.23901	1	9.8692	9.08359	1.0364×10^{-5}	1.6602×10^{-17}
1 cc atm/mole	0.024218	0.10133	1	8.470×10^{-3}	1.0501×10^{-5}	1.6823×10^{-18}
1 cm^{-1}	2.8593	11.963	118.07	1	1.2398×10^{-4}	1.9862×10^{-16}
1 eV	23,062	96,493	9.523×10^5	8065.7	1	1.2600×10^{-12}
1 erg/molecule	1.4396×10^{16}	6.023×10^{16}	5.944×10^{17}	5.0348×10^{15}	6.2422×10^{11}	1

Index

Sulfur: chemical shift of the $1s$ electron shell in atoms of, as function of oxidation state in inorganic compounds of, 183; electron affinity of, 168; electron band gap of, 129

Superheating, 46

Supersaturated vapor, 78

Surface: alloys, 42; area, 53, 54, 56, 59, 60, 61, 94, 217; area, occupied by adsorbed molecules, 217; area, relative, 218; atoms, density of, 26; atoms, dynamics of, 82; atoms, energy exchange with, 189; atoms, mean-square displacement of, 97, 98, 99, 100, 104, 119; buckling, 36, 39, 41; capacitance, 154; chemical composition, variation of, 44; chemisorption on, 255; composition of, 64; concentration, 61, 62, 63; concentration, of atoms, 4; conductance, 152, 153, 154, 155, 156, 157, 252; coverage, 212, 213, 214, 227, 244; coverage, of adsorbed molecules, 246; curved, 70, 72; Debye temperature, 210; diffraction of gas atoms by, 195; diffusion, 69, 101, 102, 110, 113, 114; diffusion, coefficients, 110, 111, 112; diffusion, mechanism, 116; distortions, 66; electrical properties of, 120, 121; emission properties of, 121; energy, specific, 52, 53, 55, 58, 59, 65, 66, 67, 68; enthalpy, specific, 53, 59, 60; enthalpy, temperature derivative of, 60; entropy, specific, 53, 58, 59, 80; films, monomolecular, 222; forces, 188; operating normal to surface, 70; forces, operating tangential to surface, 70; free energy, 58, 62, 69; free energy, excess, 62; free energy, as function of crystal orientation, 68; free energy, minimum, 68; free energy, of surface tension, 59; heat capacity, 94, 95, 97, 118; heat capacity, measurements, 60; heat capacity, ratio of bulk and,

97; heat capacity, specific, 60; heat capacity, temperature dependence of, 94; heterogeneous, model of, 102; ionization, 166, 167, 168, 189; ionization, negative, 169; lattice, expansion at, 66; layer, hexagonal, 41; melting, 46; metal, interaction potential between approaching gas atoms and, 190; net, 18; occupied by adsorbed molecules, 217; ordered domains, 34, 47; ordering, 59, 233; orientation, 46; phase transformation, 42, 44, 51, 236; phonon spectra, 209; plasmons, 174; potential barrier, 145; pressure, 22, 54; internal and external, 69, 70, 81; properties of liquids, 70; reconstruction, 44; relaxation, 36, 42; roughness, 75, 195; self-diffusion, 119; self-diffusion, active energy of diffusion for, 111, 114; semiconductor, during chemisorption, charge transfer at, 249; sites, 5, 213; space charge, 132, 145, 148; states, electronic, 132, 145, 152, 155, 156; stepped, 42; structure, 18, 19, 20, 21, 22, 23, 24, 36, 37, 40, 41, 44, 46; structure, of adsorbed gases, 9, 213, 218, 228–234; structure, amorphous, 240; structure, of chemisorbed atoms, 256; structure, coadsorbed, 238; structure, rearrangements, 39; structure, reconstructed, 239; structure, rotated, 18; structure, semiconductor, 37, 51; structure, three-dimensional, 241; structure, two-dimensional, 10; tension, 53, 55, 56, 57, 58, 59, 61, 63, 65, 69 75–76, 81, 222; tension, along different crystal faces, ratio of, 69; tension, as function of temperature, 62; tension, of ionic metal and noble gas crystals, 65; tension, of isooctane-benzene, 64; tension, of liquid, 72; tension, of liquid and solid in vacuum, 75; tension, of multicomponent sys-